Epoxide Resins

EDITOR ENGINEERING SERIES
R. M. Ogorkiewicz
M.SC.(ENG.), A.C.G.I., D.I.C., C.ENG., M.I.MECH.E.

EDITOR CHEMISTRY SERIES
Professor P. D. Ritchie
PH.D., B.SC., F.R.I.C., F.R.S.E., F.P.I.

Epoxide Resins

W. G. POTTER, B.SC., PH.D., A.R.I.C.

PUBLISHED FOR THE PLASTICS INSTITUTE

LONDON ILIFFE BOOKS

ENGLAND
Butterworth & Co (Publishers) Ltd
London: 88 Kingsway, WC2B 6AB

AUSTRALIA
Butterworth & Co (Australia) Ltd
Sydney: 20 Loftus Street
Melbourne: 343 Little Collins Street
Brisbane: 240 Queen Street

CANADA
Butterworth & Co (Canada) Ltd
Toronto: 14 Curity Avenue, 374

NEW ZEALAND
Butterworth & Co (New Zealand) Ltd
Wellington: 49/51 Ballance Street
Auckland: 35 High Street

SOUTH AFRICA
Butterworth & Co (South Africa) (Pty) Ltd
Durban: 33/35 Beach Grove

First published in 1970 for the Plastics Institute,
11, Hobart Place, London, S.W.1
by Iliffe Books, an imprint of
the Butterworth Group

ISBN 0 592 05440 3

Printed in England by J. W. Arrowsmith Ltd, Bristol

Filmset by V. Siviter Smith & Co Ltd, Birmingham

Contents

Preface

Epoxide resin consumption cannot be compared with the vast quantities of thermoplastic materials that are used annually. But in terms of versatility, breadth of application, and the complexity of the science and technology involved, epoxides are more than equal.

Any study of the resins must include the nature of the resins themselves and the ways in which they cross-link, and in addition, the innumerable possibilities in compounding and applying the resins in their many different uses. This Monograph is intended as a general introduction to the chemistry and applications of epoxide resins. The treatment although comprehensive is not exhaustive, but will it is hoped stimulate the reader into finding out more for himself. The references have been chosen with this in mind.

I am indebted to my former colleagues in the Shell Group of companies, and in particular to Tom Mika and Derek Shingleton, for the valuable help and advice they have given me.

W.G.P.

CHAPTER 1

Introduction to Epoxide Resins

1.1 BASIC CONCEPTS

Epoxide resins contain the epoxide group, also called the epoxy, oxirane, or ethoxyline group, which is a three-membered oxide ring. The simplest compound in which it is found is ethylene oxide, and Fig. 1.1 gives the formulae for a number of substituted ethylene oxides which are relevant to epoxide resin technology.

The resins can be regarded as compounds which contain, on average, more than one epoxide group per molecule; and they are polymerised through these epoxide groups, using a cross-linking

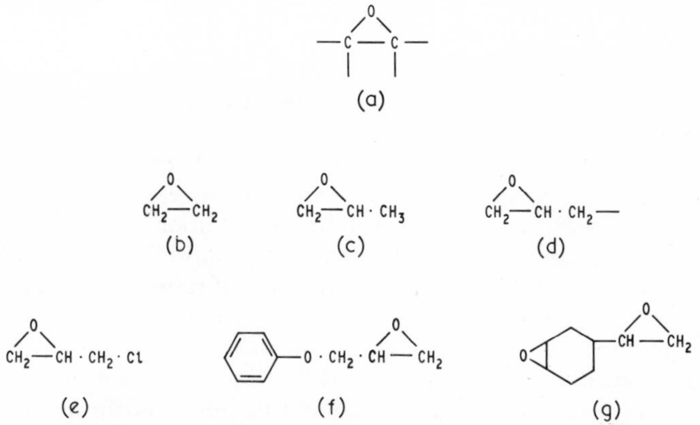

(a)

(b)   (c)   (d)

(e)   (f)   (g)

Fig. 1.1 (a) Epoxide group; (b) ethylene oxide; (c) propylene oxide; (d) glycidyl group; (e) glycidyl chloride (epichlorohydrin: ECH); (f) phenyl glycidyl ether (PGE); (g) vinyl-cyclohex-3-ene dioxide (VCDO)

agent (also called a curing agent or hardener), to form a tough three-dimensional network. It is in this cured form, when all the epoxide groups have reacted, that the resins are almost always used; in the uncured non-cross-linked state they are of limited utility.

The parent resins can be broadly classified into the following five chemical groups.

(i)   Glycidyl ethers

$$R \cdot O \cdot CH_2 \cdot \overset{\displaystyle O}{\overset{\displaystyle \triangle}{CH-CH_2}}$$

(ii)   Glycidyl esters

$$R \cdot CO_2 \cdot CH_2 \cdot \overset{\displaystyle O}{\overset{\displaystyle \triangle}{CH-CH_2}}$$

(iii)   Glycidyl amines

$$R R'N \cdot CH_2 \cdot \overset{\displaystyle O}{\overset{\displaystyle \triangle}{CH-CH_2}}$$

(iv)   Linear aliphatic

$$R \cdot \overset{\displaystyle O}{\overset{\displaystyle \triangle}{CH-CH}} \cdot R' \cdot \overset{\displaystyle O}{\overset{\displaystyle \triangle}{CH-CH}} \cdot R''$$

(v)   Cycloaliphatic

The epoxides in groups (i)–(iii) are usually prepared via a condensation reaction between the appropriate diol, dibasic acid, or diamine and epichlorohydrin (ECH) with the elimination of a simple molecule, hydrogen chloride; whereas the epoxidised olefins (groups (iv) and (v)) are formed by an addition reaction—the peroxidation of an olefinic double bond by means of a peracid such as peracetic acid. The most important group of resins commercially comprises the glycidyl ethers of dihydroxy compounds; and 95% of all epoxide resins are made by the interaction of epichlorohydrin (ECH) and diphenylolpropane (DPP). Their general formula, given below, will be discussed in detail in Chapter 2; for the moment, it need only be noted that resins with different degrees of polymerisation (i.e., different values of n) are possible.

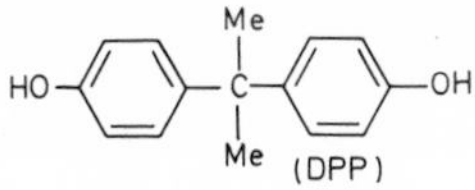

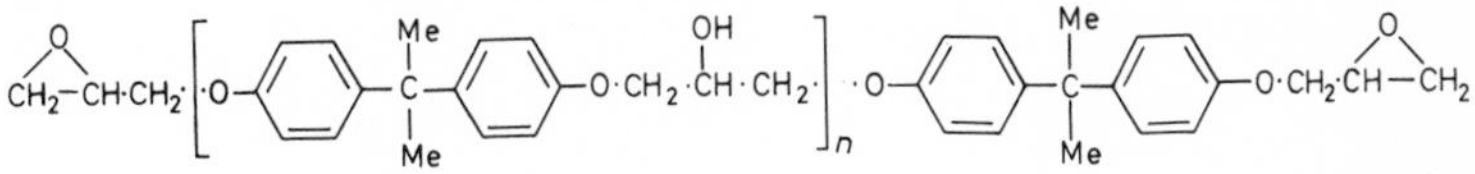

The resin corresponding to $n = 0$ in the formula is a low-melting solid which can exist as a supercooled liquid, and at values of $n > 1$ the resins are solids of increasing melting point. These are fully described in subsequent Chapters; and throughout this Monograph, unless otherwise stated, the epoxide resins discussed will be of this type. Of the remaining resin types, glycidyl esters and amines are manufactured to a very limited extent; and linear aliphatic poly-epoxides, i.e., the epoxidised olefins, have failed so far to find a significant place in the market. The cycloaliphatic diepoxides are also produced in relatively small amounts, and are used in certain specialised applications requiring that a particular property shall be outstandingly good. They are, however, important components of the resin manufacturer's range of products, and will doubtless become of increasing importance in the future.

The resins will, in general, react with compounds containing active hydrogen atoms, such as phenols, alcohols, thiols, primary and secondary amines, and carboxylic acids. The reactivity of the epoxide group towards any of these reagents will be different for each of the five resin types, depending as it does upon the electronic environment of the group and steric factors. For example, the glycidyl ethers react more rapidly with polyamines, but less rapidly than do cycloaliphatics and epoxidised olefins with the acidic reagents by means of which they can be cured (Chapter 3).

In these reactions the epoxide ring is opened to form a primary or secondary alcohol:

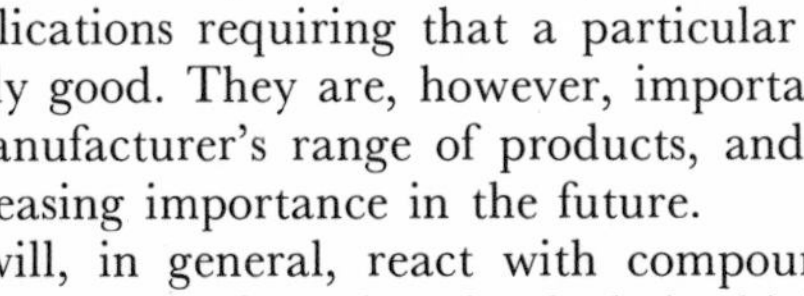

Thus, when a diepoxide (functionality 2) reacts with a substance possessing four active hydrogen atoms (functionality 4), a system of functionality 4:2 is created. This produces a cross-linked structure. The hydroxyl groups formed when the epoxide group opens may, in certain circumstances, also react with epoxide and other groups present, and in this case the effective functionality of each epoxide group increases from 1 to 2 and therefore the functionality of the diepoxide compound from 2 to 4.

A typical example of this polymerisation is the reaction between an epoxide resin and a primary diamine (Fig. 1.2).

The general formula for the glycidyl ethers of DPP shows that when $n \geqslant 1$ the resin contains secondary hydroxyls as well as epoxide groups. These hydroxyls can also join in cross-linking reactions, and in certain cases such as the reaction between a high molecular

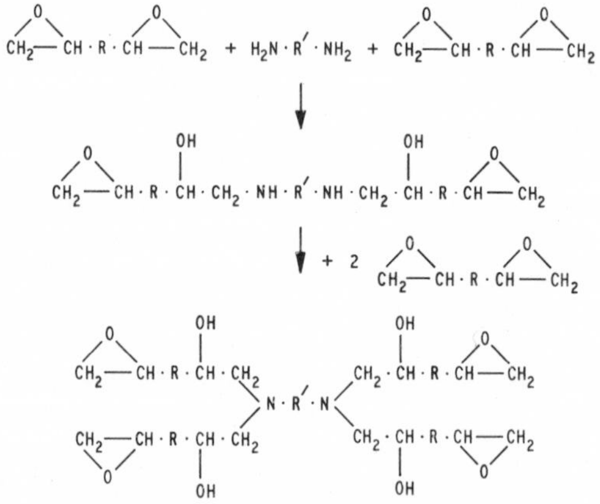

Fig. 1.2

weight epoxide and a phenolic resin, the hydroxyl reaction is the most important one involved in curing. Three-dimensional cross-linked networks are also obtained when the resins are homopolymerised through their epoxide groups, using as catalysts Lewis acids such as boron trifluoride derivatives or bases such as tertiary amines.

In practice the resin-curing agent combination often has other materials added to it, such as inert fillers, diluents, flexibilisers, and flame-retardants, which are intended to achieve certain physical or chemical properties in the cured resin, or to cheapen it. The choice of the most appropriate curing agent and other ingredients of a resin formulation is often very difficult, since the number of different possible combinations of materials is very great.

The cured resins have the well-known properties of outstanding toughness, adhesion to many substrates, chemical resistance, high mechanical strength, and high electrical resistance. Their versatility has led to many applications, including surface coatings, adhesives, monolithic flooring, laminates, castings, encapsulated electrical and electronic units, caulking and sealing compounds.

1.2 HISTORICAL DEVELOPMENTS

Work on diepoxides was mentioned in patents in the late 1920s and early 1930s, but the beginning of today's epoxide resin technology is usually taken to be the patent of Schlack of I. G. Farben[1] whose application date was December 1934. This patent, although concerned with the production of polyamines, mentioned polyglycidyl ethers of polyphenols including DPP. A little later, Castan, a Swiss chemist employed by the dental products manufacturer, De Trey Frères, of Zürich, filed two patents,[2,3] with application dates 1938 and 1943 respectively, which described the production of diglycidyl ethers and esters including a resin based on DPP and ECH (formulae as above), and polymerisation of these resins with acid anhydrides such as phthalic anhydride, and organic and inorganic bases including amines. An attempt to market Castan's products in the early 1940s as a casting resin for dental use failed, and the patents were subsequently licensed to CIBA A.G. of Basle. At the Swiss Industries Fair in 1946 this Company demonstrated the use of an epoxide resin adhesive, Araldite Type I, to bond light alloys, and at the same time offered samples of an epoxide casting resin to four Swiss electrical companies. This introduction of the resins to industry can be taken as the beginning of the commerical exploitation of these remarkable materials.

Parallel with this European activity, the paint company Devoe and Raynolds in the U.S.A. had been working with Shell Chemical Corporation to develop epoxide resins suitable for the surface coatings industry. This led to a long series of patents by Greenlee and co-workers, of Devoe and Raynolds, the first[4] being filed in September 1943. The Greenlee work covered methods of preparing DPP–ECH resins of higher molecular weight[5-8] and important ways of modifying the resins or combining them with other materials to form surface coatings. Outstanding amongst these were the esterification of epoxide resins via their hydroxyl as well as epoxide groups, to form drying or non-drying resin esters,[9] and the combination of epoxide with phenolic[10,11] or amino[12,13] resins to produce stoving finishes. Whilst this basic work was being carried out, the Shell researchers were investigating a vast range of possible curing agents for the resins, monoepoxide compounds as reactive diluents, and other possible modifications to the resin-curing agent combination.[14-19] Shell obtained licences on the Devoe and Raynolds patents and began to market a range of liquid and solid epoxide resins under the name Epon in the U.S.A. and Epikote in all other countries, principally to the surface coatings industry.

In the late 1950s and early 1960s other resin types began to appear in the manufacturers' catalogues, including epoxidised novolaks and other polyfunctional epoxides, resins derived from halogenated DPP for flame retardancy, and resins to impart flexibility to castings. Greenspan and co-workers of the Food Machinery Corporation, U.S.A., reported on resins derived from the epoxidation of polyolefins, and a range of these resins was brought to the market. The research teams of Union Carbide Corporation (Philips and co-workers) and CIBA, Basle (Batzer, Ernst, Fisch, Porrett, *et al.*) investigated very many cycloaliphatic diepoxides and market development was commenced with a small number of selected products.

Much has therefore happened since the early days of epoxide resin technology in the late 1940s and early 1950s, when CIBA and Shell first began to manufacture and market these new and versatile materials. Many more resin manufacturers have begun to produce epoxides, an extensive literature has been built up covering all aspects of their science and technology, the exploitation of their outstanding properties in an ever-growing number of uses has continued, and newer types of epoxide resins and curing agents have been developed. Further details of this historical development are available in useful reviews recently published.[20,21]

1.3 COMMERCIAL BACKGROUND

In 1968, the U.S.A. accounted for about half of the total world-wide sales of epoxide resins. European markets, including the U.K., amounted to a third of the tonnage, and the U.K. alone consumed about 8 000 tons. Production figures[22] for the U.K. show that some 7 200 tons of resin were manufactures in 1968; and outside of the U.S.A. and Europe, Japan is the next major market. Historically, the world market has grown on average by at least 10–15% per annum over the last ten years.

At least 95% of all epoxide resin sold is of the diglycidyl ether of

DPP type, an important reason being that DPP is at present the cheapest suitable aromatic dihydroxy compound available (see Section 2.1). The remaining percentage includes the cycloaliphatics, epoxidised olefins, and glycidyl ether resins such as the epoxidised novolaks, brominated DPPs, etc. An end-use analysis of the U.K. market in 1967 and 1968 is given in Table 1.1, and shows the dominant position of the surface coatings and electrical/electronic industries as consumers of epoxide resins. This situation has not changed greatly over the last ten years, and the continuous flow of new

Table 1.1 END-USE ANALYSIS[22]

	1968 %	1967 %
Surface coatings	57	53
Electrical/electronic	25	30
Adhesives	5	5
Flooring	5	4
Laminating	3	3
Tooling	3	3
Miscellaneous	2	2
Total	100	100

developments in the coating uses for the resins will ensure that the pattern does not change rapidly in the near future. The end-use pattern in the U.S.A. is similar to that in the U.K., except that surface coatings account for about 44% of total sales.

The resins are now produced in most industrialised countries of the world and the early pioneering companies of CIBA, Devoe and Raynolds, and Shell have been joined by Dow and Union Carbide, who are also producers of one or both of the raw materials DPP and ECH, and many other large and small European and American chemical manufacturers. At present there are 14 producers of epoxide resins in Europe alone.

The price of DPP/ECH resins varies from country to country but as a general guide the price of the liquid grades in the U.K. in 1968 was about £660 per ton and solid grades £560 per ton. U.S. prices are significantly below these levels. For comparison, U.K. prices for polyesters can be regarded as £200–£250 per ton.

REFERENCES

1. SCHLACK, P., Ger. Pat. 676,117
2. CASTAN, P., Brit. Pat. 518,057
3. CASTAN, P., Brit. Pat. 579,698
4. GREENLEE, S. O., U.S. Pat. 2,503,726
5. GREENLEE, S. O., U.S. Pat. 2,582,985
6. GREENLEE, S. O., U.S. Pat. 2,615,007
7. GREENLEE, S. O., U.S. Pat. 2,615,008
8. GREENLEE, S. O., U.S. Pat. 2,694,694
9. GREENLEE, S. O., U.S. Pat. 2,456,408
10. GREENLEE, S. O., U.S. Pat. 2,521,911
11. GREENLEE, S. O., U.S. Pat. 2,521,912
12. GREENLEE, S. O., U.S. Pat. 2,528,360
13. GREENLEE, S. O., U.S. Pat. 2,528,359
14. NEWEY, H. A., U.S. Pat. 2,553,718
15. NEWEY, H. A., U.S. Pat. 2,575,538
16. BRADLEY, T. F., U.S. Pat. 2,500,449
17. NEWEY, H. A. and WILES, Q. T., U.S. Pat. 2,528,932
18. WILES, Q. T., U.S. Pat. 2,528,933
19. WILES, Q. T., U.S. Pat. 2,528,934
20. ANON., *Appl. Plast.*, **10,** No. 3, 40 (1967)
21. PREISWERK, E., *Technica*, **4,** 247 (1965) and **5,** 355 (1965)
22. ANON., *Br. Plast.*, **71** (1969)

CHAPTER 2

Synthesis of Resins containing the Glycidyl Group

2.1 GLYCIDYL ETHERS

The epoxide resins first developed commercially and still completely dominating world-wide markets are those based on diphenylolpropane (DPP), also known as Bisphenol A, and epichlorohydrin (ECH). The polyfunctional phenol (DPP) and the epoxide-containing compound (ECH) are condensed together in the presence of aqueous caustic soda at atmospheric pressure and slightly elevated temperatures. Heat is evolved and sodium chloride and water formed, together with the epoxide resin. This basic process was described by Castan,[1,2] and its further development, chiefly by resin manufacturers, is revealed by the extensive patent literature on the subject.

The resins can be represented by the general formula in Section 1.1,

from which it can be seen that the molecular species concerned is a linear polyether with terminal epoxide groups and secondary hydroxyl groups occurring at regular intervals along the length of the macromolecule, a large part of the backbone of the chain consisting of aromatic rings.

The number of repeating units (n) is dependent essentially on the molar ratio of ECH:DPP used during the resin synthesis. A series of resins may therefore be obtained, each grade consisting of a mixture of molecules differing in chain length and molecular weight (M). Resins of short chain length and low M are syrupy viscous liquids, whereas resins of longer chain length and higher M are hard, brittle, amber-coloured solids at normal temperatures. Usually, an individual resin is characterised by its melting point, viscosity, and equivalent weight per epoxide group (WPE), these properties being indicative of average chain length and number-average molecular weight M_n. Theoretical values of these properties for a range of resins $(n = 0$ to $n = 7)$ are shown in Table 2.1.

Table 2.1 THEORETICAL VALUES OF SOME EPOXIDE RESIN PROPERTIES[33]

Value of n	Molar ratio ECH:DPP	Molecular weight M_n	No. of epoxide groups	No. of hydroxyl groups	Epoxide equivalent WPE
0	2:1	340	2	0	170
1	3:2 (1·5:1)	624	2	1	312
2	4:3 (1·3:1)	908	2	2	452
3	5:4 (1·25:1)	1 192	2	3	596
4	6:5 (1·2:1)	1 476	2	4	738
5	7:6 (1·16:1)	1 760	2	5	880
6	8:7 (1·14:1)	2 044	2	6	1 022
7	9:8 (1·17:1)	2 328	2	7	1 164

The values given in Table 2.1 are not those found for actual resins, since the theoretical values assume the resin to be composed entirely of one molecular species, the linear diepoxide with the appropriate value of n. However, this is not the true situation. Except for specially prepared batches, commercial resins consist of a mixture of molecules having different chain lengths, some not even possessing two terminal epoxide groups. The complex series of reactions that can occur in the reactor during resin synthesis also leads to some molecules having one terminal phenolic hydroxyl group or chlorohydrin group. Actual values for a series of resins are given in Table 2.2.

The chief reasons for the development of resins based on DPP and ECH, and also for their continued wide use, were (a) the ready availability of a relatively cheap dihydric phenol (DPP) from

Table 2.2 ACTUAL VALUES OF SOME EPOXIDE RESIN PROPERTIES[34]

Value of n	Molar ratio ECH:DPP	Molecular weight M_n	No. of epoxide groups	No. of hydroxyl groups	Epoxide equivalent WPE
0·0	10:1*	380	—	—	175–210
2·0	1·57:1	900	—	—	450–525
3·7	1·22:1	1 400	—	—	870–1 025
8·8	1·15:1	2 900	—	—	1 650–2 050
12·0	1·11:1	3 750	—	—	2 400–4 000

*See Section 2.1.3 for reason for this ratio

petroleum raw materials, and (*b*) the good mechanical and chemical properties exhibited by these resins when hardened by a large number of different cross-linking agents.

2.1.1 DIPHENYLOLPROPANE (DPP)

This dihydric phenol is readily prepared (m.p. 155 °C) by allowing acetone to react with an excess of phenol at slightly elevated temperatures (50 °C) in the presence of a strong acid catalyst such as 75% sulphuric acid or gaseous hydrogen chloride (Fig. 2.1):

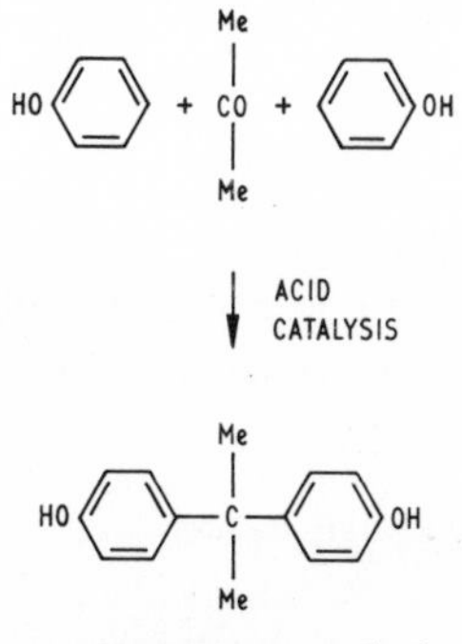

DIPHENYLOLPROPANE (DPP)

Fig. 2.1

The usual batch process can yield flake DPP of very high purity. For epoxide resin manufacture, a *p,p*′-isomer content of 98% minimum is usually required, although certain producers use a product containing 95% *p,p*′-content. There is no difficulty in obtaining the 98% quality, the remaining 2% being composed of

o,p'- and o,o'-isomers. DPP is also widely used for polycarbonate manufacture, and here a much higher level of purity is needed, as regards not only isomer content but also impurities such as iron, arsenic, and certain coloured organic molecules. The development and persistence of colour in epoxide resins can be partly attributed to the presence of iron and coloured organic impurities in the DPP.

2.1.2 EPICHLOROHYDRIN (ECH)

Originally prepared from glycerol, epichlorohydrin is now produced as an alternative product in the synthetic route to glycerol starting from petroleum raw materials. Propylene from petroleum cracking operations is allowed to react with chlorine gas at high temperatures and slight pressure. Under these conditions allyl

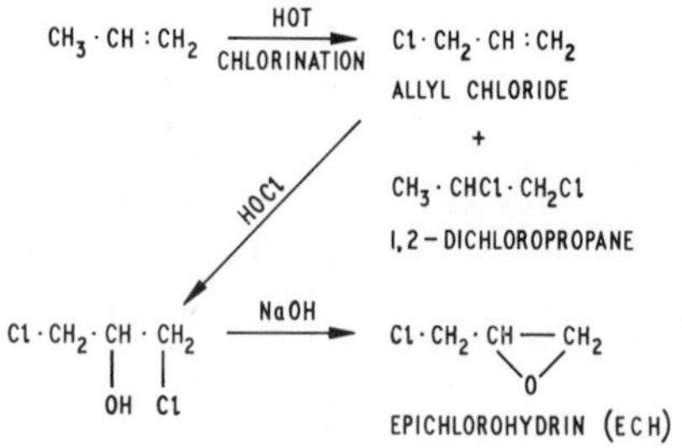

Fig. 2.2

chloride is obtained as well as the more predictable product 1,2-dichloropropane. The allyl chloride is then treated with dilute hypochlorous acid, forming glycerol dichlorohydrin, which is dehydrochlorinated by using slaked lime or caustic soda (Fig. 2.2).

2.1.3 LOW MOLECULAR WEIGHT RESINS

The simplest member of the series of glycidyl ether resins based on DPP and ECH is the following diglycidyl ether of DPP ($n=0$ in the general formula, Section 1.1).

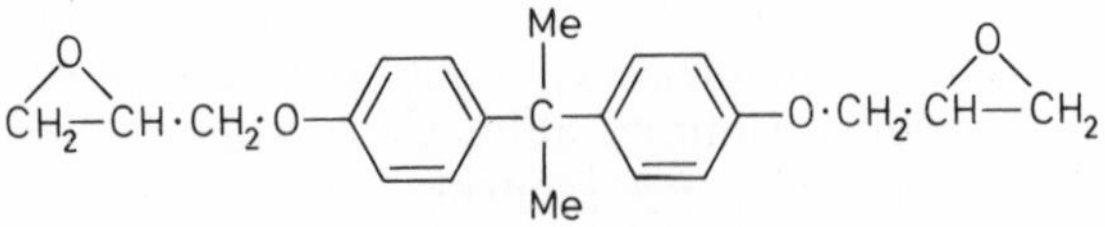

Diglycidyl ether of DPP

This resin has a theoretical M of 350, two terminal epoxide groups, and no hydroxyl groups. In principle, it is prepared by condensing ECH with DPP in the presence of aqueous caustic soda, the theoretical ratio of ECH to DPP being 2:1. The reactions involved in the synthesis are shown in Fig. 2.3. In stage (a) the epoxide group of the ECH reacts with the phenolic hydroxyl, probably as the phenolate ion, under the catalytic influence of the alkali, forming the chlorohydrin ether. At stage (b) the chlorohydrin is dehydrochlorinated by the alkali, forming the glycidyl ether together with sodium chloride and water. The unreacted phenolic hydroxyl then reacts with a second molecule of ECH (stage (c)) repeating stages (a) and (b), i.e. chlorohydrin formation followed by dehydrochlorination to form an epoxide group. In this way the diglycidyl ether of DPP is formed. There is, however, a competing reaction for DPP (stage (d)) which leads to higher molecular weight diphenols which can then react further with ECH. In order to suppress the formation of this higher species, a large excess of ECH is used over the stoichiometric amount of 2 moles. In fact a ratio of ECH:DPP of 2:1 leads to the formation of only 10% of the diglycidyl ether of DPP. Therefore, in practice, minimum ratios of ECH:DPP of 10:1 are employed, the ECH acting as a reaction solvent as well as a reactant.[3] The reactor must contain sufficient ECH to ensure that all the phenolic hydroxyls are consumed by reactions (a) and (c), leaving little or no DPP available for reaction (d).

Goppel[4] has described the need to maintain the water content of the reaction mixture between 0·3% and 2% by weight throughout the reaction, if high yields (90–95%) are to be obtained. This is achieved by distilling off the water as an azeotrope with the ECH and returning the ECH to the reactor after separation of the water. The reaction mixture can also contain an inert solvent such as toluene or xylene, which helps the separation of the azeotrope and salt removal at the working-up stage.[5] No reaction occurs under anhydrous conditions, and undesired byproducts are formed if the water content is greater than 2%. The water content can also be partly controlled by the rate of addition of the caustic soda solution.

The sodium hydroxide used in the process has a double role to play:
(*a*) As a catalyst for the epoxide/phenolic hydroxyl reaction.
(*b*) As the means of dehydrochlorination.
The use of alternative catalysts such as lithium salts[6] or quarternary ammonium compounds in stage (a) (Fig. 2.3) of the reaction process (formation of the chlorohydrin ether from the phenolic hydroxyl/

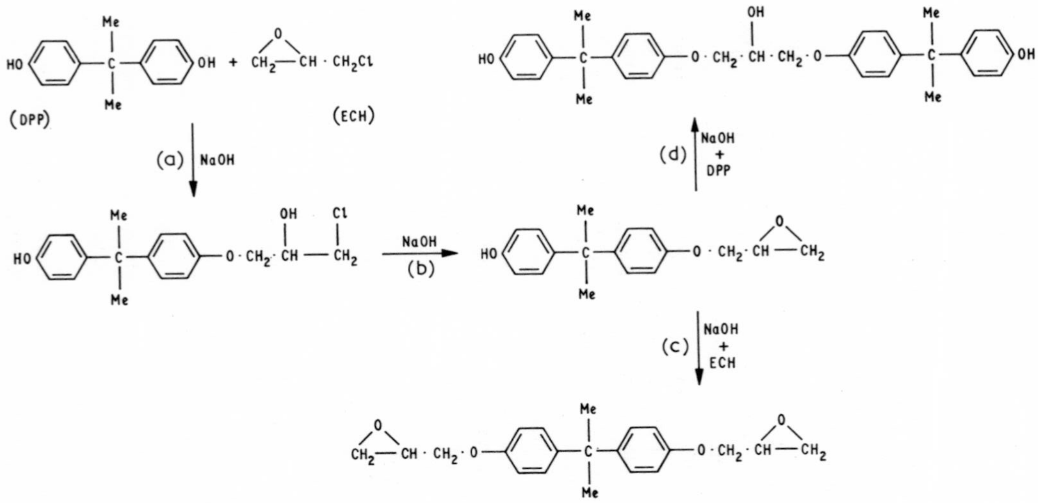

Fig. 2.3

epoxide reaction) has been claimed. Stage (b) of the process (dehydrochlorination) would use sodium hydroxide as previously described.

The manufacture of the low-*M* liquid resins is carried out as a batch process in stainless steel kettles equipped with powerful

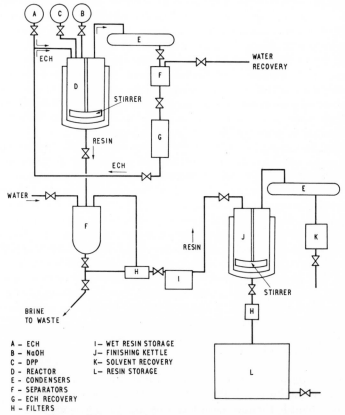

A – ECH
B – NaOH
C – DPP
D – REACTOR
E – CONDENSERS
F – SEPARATORS
G – ECH RECOVERY
H – FILTERS

I– WET RESIN STORAGE
J– FINISHING KETTLE
K– SOLVENT RECOVERY
L– RESIN STORAGE

Fig. 2.4 Typical plant layout for liquid epoxide resin production

anchor stirrers. Fig. 2.4 shows a typical plant layout and flow diagram for the production of liquid epoxide resins. In a production sequence, the ECH (10 moles) is added to the kettle, followed by the DPP (1 mole). The water content of the mixture is then reduced to below 2% by heating the mixture until the ECH/water azeotrope distils off. The ECH layer, after condensation, is returned to the kettle, the water being discarded. When the necessary water content has been reached, the reaction is started by the slow addition of 40% sodium hydroxide solution, the temperature being maintained at about 100 °C.

When all the alkali is added, which takes about 2–3 hr, the excess ECH is recovered by distillation at reduced pressure. The crude reaction product is cooled to 120 °C and a solvent such as toluene or methyl isobutyl ketone added to dissolve the resin and leave the salt formed in the reaction as a slurry.

Hot water is added to wash the resin solution and remove the brine. After filtration and further washing, the solvent is removed by distillation and the resin dried by heating under vacuum or by the use of a thin film evaporation system. It is then filtered into storage tanks, where it is frequently kept for considerable periods at temperatures of 60 °C without any deterioration.

The pure diglycidyl ether of DPP is a solid (m.p. 43 °C), whereas the commercial grades of the resin which contain a proportion of the high-M materials are 'supercooled liquids' having viscosities of 100–140 poises at room temperature. An Appendix to this Chapter gives the characteristics of some typical commercial liquid resins. However, the crystallisation of liquid epoxide resins can cause practical difficulties, particularly when the resin is part of a formulated product. As would be expected, the diglycidyl ether of DPP tends to crystallise out as resins increase in purity and particularly if they are kept under cold storage conditions and crystallisation nuclei are present such as filler particles. Simple warming to 40 °C will usually be sufficient to restore a resin to its previous liquid form.

2.1.4 HIGHER MOLECULAR WEIGHT RESINS

The high-M glycidyl ethers of DPP usually manufactured have values of n in the general formula (Section 2.1) ranging from 2 to 12. If no chain-branching occurs during the preparation, each resin should theoretically have a maximum of two terminal epoxide groups per molecule and as many secondary hydroxyl groups as there are repeating units in the molecule, corresponding to the value of n. These resins are synthesised similarly to the low-M types, by allowing ECH and DPP to interact in the presence of excess sodium hydroxide. Variation of the ECH:DPP and ECH:NaOH ratios enables resins with different molecular weights to be obtained, and Table 2.2 gives actual values for the ratios that have been employed to prepare specific resins. Theoretically, to obtain a resin with a given value of n for the repeating unit, $(n+1)$ moles of DPP should be condensed with $(n+2)$ moles of ECH using $(n+2)$ moles of NaOH. This would yield 1 mole of the resin plus $(n+2)$ moles of water and $(n+2)$ moles of NaCl.

Much of the original development work on the synthesis of the higher molecular weight resins was carried out by Greenlee, and is described in the patent literature.[7-11]

The mechanism of the synthesis is similar to that described previously for the liquid grades, but here the reduced ratio of ECH to DPP allows reaction (d) in Fig. 2.3 to take place. This is necessary to build up the molecular chains to the degree of polymerisation required.

Resins having up to four DPP residues in the molecule (WPE up to *c.* 1 000) are usually produced by the 'taffy' process. This consists of placing the ECH, DPP, and NaOH solution in a large resin kettle equipped with a powerful stirrer. Heat is supplied and the reaction sets in at about 50 °C, depending on the batch size. The considerable heat evolved during the condensation causes the temperature to rise to about 100 °C, the reaction being completed in about 1–2 hr. As the reaction proceeds a white putty-like 'taffy' is formed which is an emulsion of water in the resin and also contains salt and sodium hydroxide. This emulsion is coagulated, separated, and washed with water until free from alkali and salt. The resin is then dried by heating and stirring at 140 °C, poured into cooling pans, and subsequently crushed and bagged.

An alternative method of washing the resin uses a solvent such as methyl isobutyl ketone to dissolve the crude 'taffy'. This solution is then washed with water, the solvent removed, and the dried resin bagged.

Resins having five or more DPP residues can either be produced by the 'taffy' method described, using solvent washing, or by a two-stage method.[9] This consists of fusing a resin of lower M with more DPP at about 190 °C. The residual base content of the resin is usually sufficient to catalyse the second stage reaction between the phenolic hydroxyl and the epoxide group. In practice, the starting resin is still molten in the kettle but washed and at the end of the drying stage. The additional DPP is then added and the temperature maintained at 190 °C until the reaction is complete as judged by the viscosity of the resin.

Some typical properties of selected solid resins that are available commercially are given in an Appendix to this Chapter.

2.1.5 POSSIBLE SIDE-REACTIONS OCCURRING IN THE CONDENSATION

In practice, the resins do not consist solely of molecules of the type shown in the idealised formula, and can contain small amounts of different structures formed by side-reactions, and also trace amounts of inorganic ions. Some of these reactions are:

(*a*) Abnormal addition of the phenolate anion to the ECH epoxide group (Fig. 2.5).

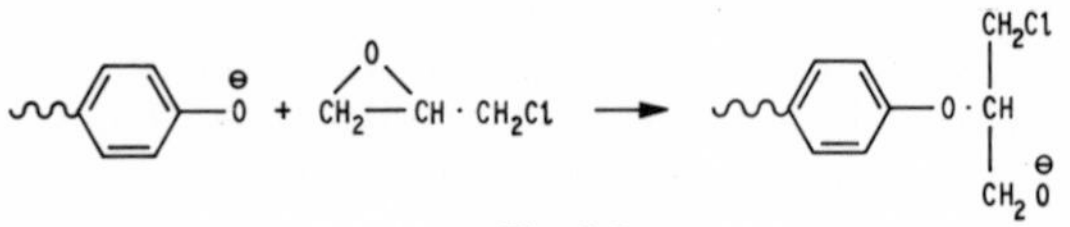

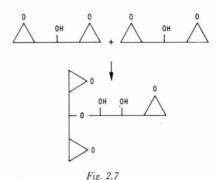

Fig. 2.5

(*b*) Hydrolysis of the epoxide group of a resin forming an *a*-glycol (Fig. 2.6).

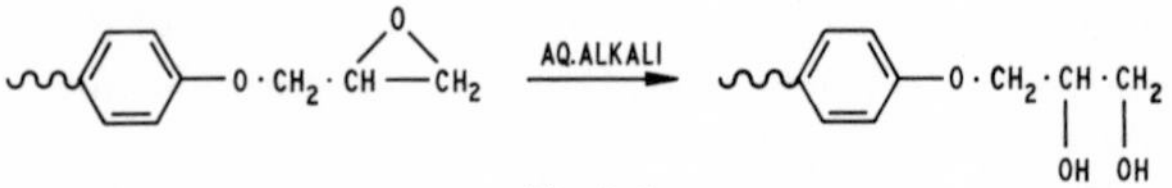

Fig. 2.6

(*c*) The *ortho*-DPP isomer is less reactive than the *para*-isomer, and is therefore a possible source of unreacted phenolic hydroxyl in the resin.

(*d*) A possible cause of chain-branching is the reaction between an epoxide group of a resin and a secondary hydroxyl group of another, this being favoured by temperatures approaching 200 °C and the presence of caustic soda, as indicated by the convenient skeleton formulae shown in Fig. 2.7.

Fig. 2.7

The use of solvents in the final washing step of resin manufacture has ensured that only trace amounts of inorganic ions are present in the resin. However, the total absence of this residual material would no doubt affect the readiness with which the resins undergo subsequent reactions such as esterification or copolymerisation with phenolic and urea resins. The removal of volatile materials from the resins is also important, only a small amount (1%) being needed to cause a significant reduction in viscosity.

2.1.6 HIGH MOLECULAR WEIGHT LINEAR POLYETHERS

The condensation of DPP with ECH has been used to produce high-M linear polyethers in which values of n in the general formula as high as c. 100 have been obtained (M_n c. 30 000). These phenoxy compounds need not be epoxide-terminated, are some 10–30 times higher in molecular weight than the ordinary solid resins, are essentially linear molecules, and are not normally used in conjunction with curing agents to achieve network structures. They do, however, contain free hydroxyl groups and these can serve as points for chemical modification of the polymer to produce cross-linked polymers, by reaction with isocyanates or melamine-formaldehyde resins. The 'phenoxies' are in fact to be regarded as high-M thermoplastic materials and have been used either as surface-coating binder resins which harden on evaporation of the solvents, or as thermoplastics for blow moulding, injection moulding, extrusion, and adhesives.

2.1.7 GLYCIDYL ETHERS OF NOVOLAK RESINS

Since the beginning of epoxide resin technology, constant effort has been devoted to obtaining resin systems which maintain their properties at higher temperatures. One approach to this problem has been to increase the cross-link density in the cured polymer network, either by the use of polyfunctional resins having average epoxide contents greater than 2 or by using curing agents of higher functionality.

A large number of polyhydroxy phenolic compounds have been used as starting points for resin synthesis. In 1956, Partansky and Schrader[12] compared the thermal stability of conventional DPP–ECH resins with an epoxidised novolak and found that in all systems tested the epoxidised novolak gave a higher heat deflection temperature (HDT) than those of the corresponding castings from the DPP–ECH resins. This was followed in 1957 by Howe *et al.*, who referred to the preparation and properties of a polyepoxide resin derived from a phenolic resin as the source of hydroxyl groups.[13] The value of a phenolic resin backbone to provide short-term high temperature stability, together with the chemical versatility of the epoxide group, was therefore recognised and led to commercial resins of this type becoming available. The properties of such a resin have been described by Applegath *et al.*[14] and in resin producers' technical literature.[15]

The epoxidised novolak resin most used in practice has 3·6 epoxide groups per molecule ($n = 1·6$ in the general formula below). The resin is synthesised by first preparing a low-M novolak from phenol and formaldehyde in the presence of concentrated sulphuric acid.[16] The novolak is then made to react with ECH in the presence of a basic catalyst such as sodium hydroxide, producing the glycidyl ether in the usual manner. Care is taken that no unreacted phenolic groups remain in the resin since this would limit storage stability and cause volatiles to be evolved during cure.

Epoxidised novolak resin

A typical epoxidised novolak has M 650, WPE 175–182, and the extremely high viscosity of 190 000 poises at 25 °C which reduces to 4–5 poises at 100 °C. To overcome handling difficulty due to the high viscosity, epoxidised novolaks are often used in solution in acetone or methyl ethyl ketone, or as a blend with a low viscosity DPP-based glycidyl ether resin, or in conjunction with a reactive diluent. Alternatively the resin can be heated to 60–70 °C, where its viscosity is around 40 poises, and it is therefore sufficiently fluid to allow easy mixing with curing agents.

Novolaks based on alkyl-substituted or other phenols can also be used to prepare epoxide resins, and certain of these can be made soluble in aliphatic hydrocarbons.[17]

2.1.8 GLYCIDYL ETHERS OF OTHER POLYHYDRIC PHENOLS

A very large number of polyhydric phenols, many of them modified DPPs, have been used to prepare diglycidyl ethers. The most important are given in Fig. 2.8.

Epoxide resins based on resorcinol (Fig. 2.8 (a)) have been developed to a commercial stage, particularly in the U.S.A.,[18] but they

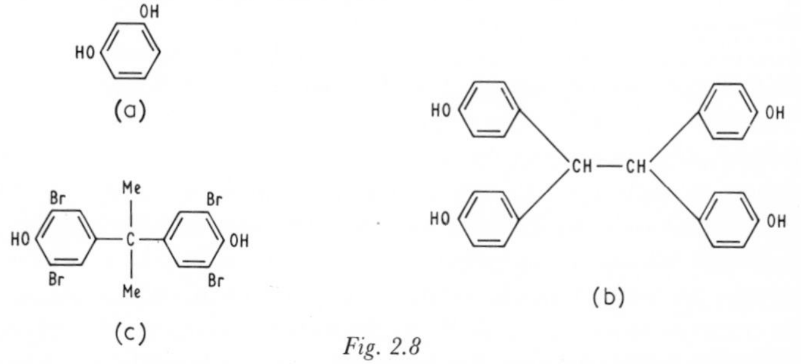

Fig. 2.8

have not established a place for themselves in the market since they do not possess any obvious advantages over the DPP-based resins.

The polyphenol 1:1,2:2-(*p*-hydroxyphenol)ethane (Fig. 2.8 (b)) is used to prepare a tetrafunctional epoxide resin,[19] aimed at providing good high temperature performance through increased ·cross-link density. The resin (m.p. 80°C) is usually employed as a blend with a low molecular weight liquid resin, or in solution just as with the epoxidised novolaks. The diglycidyl ether of tetrabromo-DPP (Fig. 2.8 (c)) is the resin most used to obtain flame retardancy in epoxide resin systems. It is a solid at room temperature and is used in conjunction with DPP–ECH resins. For its use as a flame retardent resin, see Section 5.5.

2.1.9 GLYCIDYL ETHERS OF ALIPHATIC POLYOLS

Commercial grades of glycidyl ether epoxides are available, derived from aliphatic polyols such as glycerol (a), the polyglycols (b), or pentaerythritol (c):

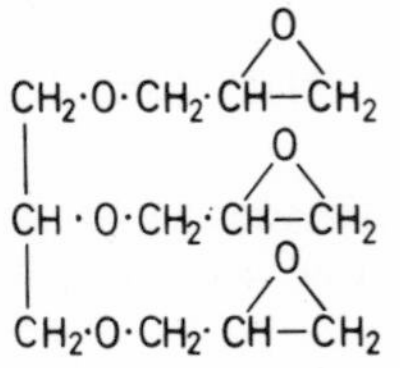

$$CH_2-CH\cdot CH_2 O\cdot \left[\cdot CH_2\cdot \overset{R}{CH}\cdot O\cdot \right]_{\underline{n}} \cdot CH_2\cdot \overset{R'}{CH}\cdot O\cdot CH_2\cdot CH-CH_2 \quad \text{(b)}$$

$$\begin{array}{c} HO\cdot CH_2 \\ \\ HO\cdot CH_2 \end{array} \Big\rangle C \Big\langle \begin{array}{c} CH_2\cdot O\cdot CH_2\cdot CH-CH_2 \\ \\ CH_2\cdot O\cdot CH_2\cdot CH-CH_2 \end{array} \quad \text{(c)}$$

They are produced similarly to the phenolic-based glycidyl ethers, by reaction of ECH with the polyol in the presence of a catalyst. However, the intermediate aliphatic chlorohydrin formed in the reaction is much more alkali-sensitive than the corresponding aromatic chlorohydrin, and can easily undergo hydrolysis to the glycol. In addition, the presence of strong alkali can cause the aliphatic epoxide, when formed, to polymerise. This has led to the development of catalysts such as aluminates, silicates, or zincates, carrying out the dehydrohalogenation[20] in a non-aqueous system, the first step (formation of the chlorohydrin) being catalysed by a Friedel-Crafts catalyst such as BF_3, $AlCl_3$, or H_2SO_4.

The aluminates apparently will remove the elements of hydrogen halide from the halohydrin but will not assist hydrolysis or polymerisation of the system in non-aqueous media such as benzene or acetone.

The resins based on glycerol and pentaerythritol are water-soluble and have low viscosities, 90–150 and 1 000–1 500 centipoises at 25 °C respectively. They can be cured with amines or anhydrides, and are more reactive towards amines than the glycidyl ethers of DPP. The pentaerythritol resin is reported to have a functionality of 2·2, to cure between two and eight times as fast as the diglycidyl ether of DPP, and to reduce by 50% the viscosity of the latter resin when used at 20% concentration.[21] It will also adhere to wet surfaces and shows excellent adhesive properties.

The polyglycol diepoxides are used as flexibilisers and are mentioned in Chapter 6.

2.2 GLYCIDYL ESTERS

The reaction between epichlorozydrin and a carboxylic acid leads to the formation of a glycidyl ester, via the intermediate chlorohydrin (Fig. 2.9).

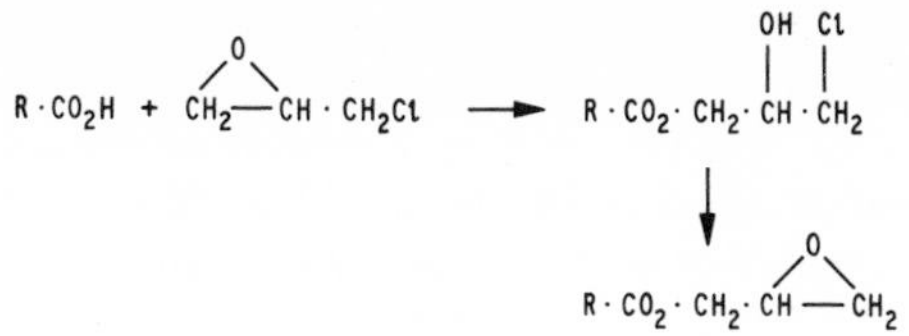

$$R \cdot CO_2H \; + \; CH_2\!\!-\!\!CH \cdot CH_2Cl \; \longrightarrow \; R \cdot CO_2 \cdot CH_2 \cdot \overset{OH}{\underset{}{C}}H \cdot \overset{Cl}{\underset{}{C}}H_2$$

$$\downarrow$$

$$R \cdot CO_2 \cdot CH_2 \cdot CH\!\!-\!\!CH_2$$

Fig. 2.9

The reaction is catalysed by alkalis, although quaternary ammonium salts[22] and tributylamine[23] have been used to bring about the first step.

A great number of acids have been made to undergo this reaction, but only the glycidyl esters derived from phthalic, isophthalic, and terephthalic acids,[24] and tetra- and hexa-hydrophthalic acids have found any significant practical applications in epoxide resin technology.

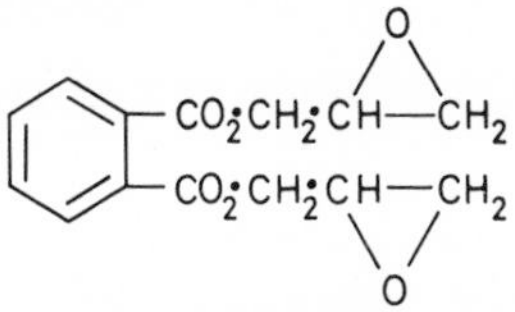

Diglycidyl phthalate 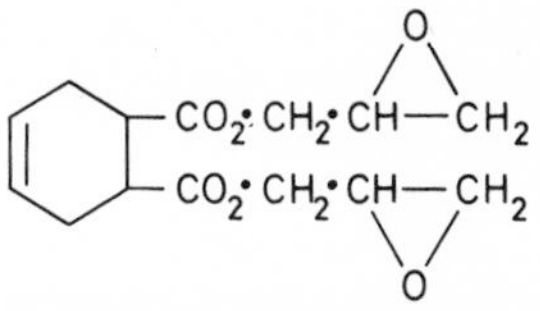   Diglycidyl tetrahydrophthalate

These esters have reactivities towards curing agents similar to those of the glycidyl ethers rather than of the cycloaliphatics.

2.3 GLYCIDYLAMINES

Glycidylamines can be prepared from aliphatic and aromatic primary and secondary amines, although the only ones of practical value to date are based on aniline and diaminodiphenylmethane (Fig. 2.10).

The patent literature describes the preparation of glycidylamines by dropwise addition of amine to the ECH, followed by the slow addition of aqueous NaOH at room temperature.[25] The epoxides are not hydrolysed by the alkali, and can be separated by distillation. The use of an inert diluent such as benzene or toluene in the reaction mixture is often an advantage.

One preparation of diglycidylaniline is carried out as follows.[25] To a solution of 4·3 moles ECH in 400 ml of boiling methanol is added 2 moles of aniline, dropwise over 15 min. The mixture is then boiled for 3 hr under reflux. The alcohol is distilled off, and 200 ml of benzene added followed by 600 ml of 44% NaOH, dropwise over 30 min at 20–25 °C. Vigorous stirring is continued for a further 3 hr and water added to dissolve the precipitated NaCl. The benzene layer is then separated from the aqueous one, dried, and distilled.

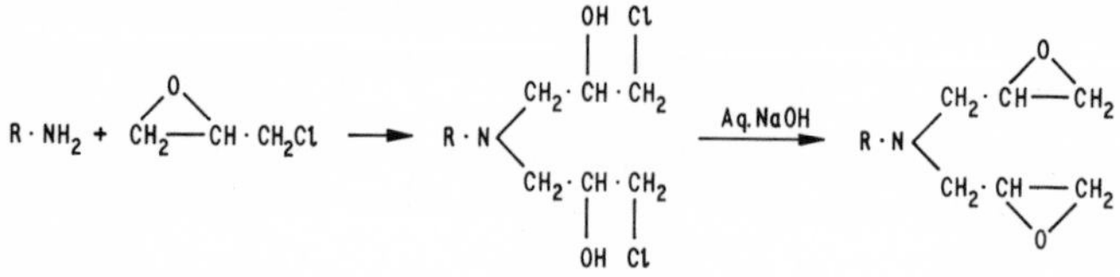

Fig. 2.10

Diglycidylaniline is obtained as a light yellow mobile liquid, b.p. 136–138 °C (yield 76–82%).

High-*M* glycidylamines have also been described. In general, glycidylamines are less reactive towards amines than glycidyl ethers, but more reactive towards anhydrides.[26]

Considerable work has been devoted to the synthesis of triglycidyl isocyanurate.[27–32]

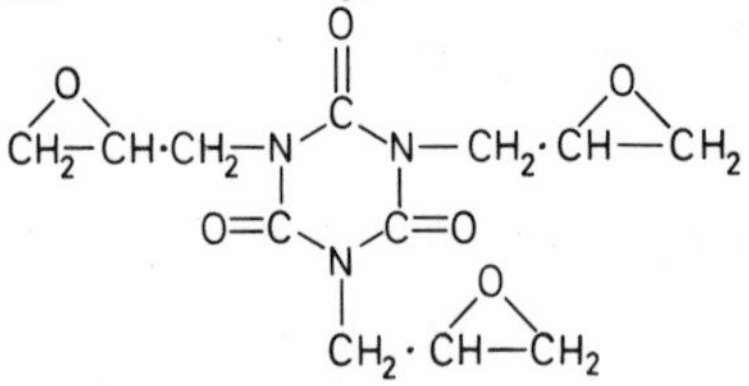

This triepoxide is synthesised from the reaction of ECH with cyanuric acid, the halohydrin formed being dehydrohalogenated with caustic soda. Cyanuric acid, which exhibits keto-enol tauto-merism, reacts in the keto form in this situation (Fig. 2.11).

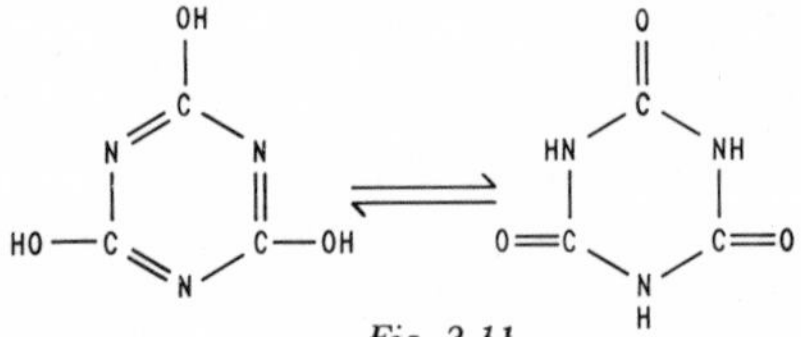

Fig. 2.11

2.4 CHARACTERISTICS OF THE UNCURED RESINS

In previous Sections it has been noted that most commercial grades of epoxide resins consist of a mixture of molecules having different degrees of polymerisation (different values of n in the general formula). Most but not all of these molecules have two terminal epoxide groups, and are usually linear unbranched structures. However, a proportion of the molecules will be branched and a very small number will be terminated at least at one end of the chain by a chlorohydrin group. Clearly, factors such as molecular weight distribution, degree of branching, and average functionality will determine the characteristics of the uncured resins and the performance of the resins in their cured state.

2.4.1 MOLECULAR WEIGHT DISTRIBUTION

Little work has been published on the determination of the molecular weight distribution of epoxide resins. Conventional fractional precipitation has been employed and also the technique of thin layer chromatography (TLC). Spell and Eddy[35] describe the use of TLC to fractionate diglycidyl ether resins and epoxidised novolaks followed by the use of infra-red spectroscopy to determine molecular weights via the concentration of functional groups in each fraction. Weatherhead[36] also used TLC to separate commercial grades of resin into their separate components. Liquid, solid, and brominated grades were examined together with an epoxidised novolak and a tetrafunctional epoxide. With the liquid grades, monomer ($n = 0$) contents of 85–95%, 80–85%, and 70–75% were obtained.

The most powerful and rapid technique for resin fractionation and molecular weight distribution determination is undoubtedly gel permeation chromatography (GPC). This technique uses a column packed with particles of an inert cross-linked gel network. The polymer sample is dissolved in a solvent whose polarity is similar to that of the gel, usually tetrahydrofuran, and is introduced into the column. The column is eluted with the same solvent and the extent to which each molecular species is able to diffuse into the gel network determines its elution time. Polymer concentration in the outflowing solution is determined by the difference in refractive index between the polymer solution and the solvent. This technique has been fully described by Moore,[37] Edwards,[38] and others.[39,40]

A series of epoxide resins has been examined by Miles[41] using GPC (Fig. 2.12). The curve for purified diglycidyl ether of DPP (Fig. 2.12(a)) shows the absence of higher molecular weight homologues, whereas the curve for a liquid resin having an epoxide

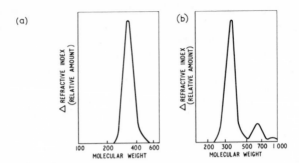

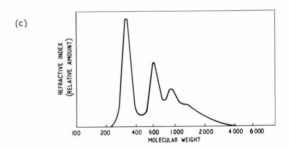

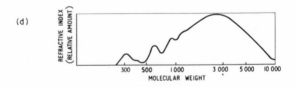

Fig. 2.12 (a) Chromatogram of purified diglycidyl ether of DPP; (b) Chromatogram of liquid epoxide resin (epoxide equivalent weight—189); (c) Chromatogram of high viscosity liquid epoxide resin (epoxide equivalent weight—274); (d) Chromatogram of solid epoxide resin (epoxide equivalent weight—982). (Courtesy: B. H. Miles; American Chemical Society Symposium)

(e)

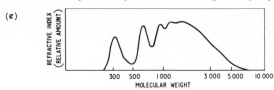

(f)

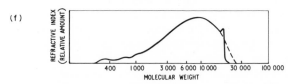

*Fig. 2.12 (e) Chromatogram of solid epoxide resin (epoxide equivalent weight—540);
(f) Chromatogram of solid epoxide resin (epoxide equivalent weight—2430) (Courtesy:
B. H. Miles; American Chemical Society Symposium)*

equivalent of 189 (Fig. 2.12(b)) indicates the distribution of homologues to be $n = 0$, $87·2\%$; $n = 1$, $11·1\%$; $n = 2$, $1·5\%$.

Curves (c)–(f) depict respectively the chromatograms of a high
viscosity liquid resin, and three solid grades, of epoxide equivalent
540, 982, and 2 430. The distribution of the various homologues is
readily calculated from the areas under the curves.

2.4.2 RESIN STRUCTURE FROM INFRA-RED SPECTRA

Infra-red absorption spectra can be used to determine (*a*) the
structure of epoxide resins and their curing agents, (*b*) the constituents of the resin system (e.g., the nature of a diluent), and (*c*) the
epoxide content of a resin sample. This technique can also be used to
determine the residual epoxide content of a cured system (due to
unconverted epoxide groups) and also to throw light upon its
structure (Chapter 5).

Examination of numerous molecules containing the epoxide
group, including the major resin types considered in this Monograph, have led to the assignment of specific absorption bands to the
epoxide group in different environments.[42–44] Characteristic spectra
for a number of epoxide resins and curing agents, and the use of these
spectra in quality control and identification of components of resin
blends, have been described by Lee and Vincent.[45] The most
comprehensive published set of infra-red spectra of uncured and
cured epoxide resins is given by Lee and Neville.[46]

In a related series of resins such as those based on DPP and ECH, the intensity of absorption of infra-red light by the various groups in the molecule increases as their relative concentration increases. This effect can therefore be used to measure the concentration of epoxide and hydroxyl groups in samples of the various resins. Kagarise and Weinberger[47] related the changes in absorption in a series of liquid and solid resins to the change in molecular weight and also to the epoxide equivalent of the resins. Dannenberg and Harp[48] also used the same technique, and Dannenberg[49] extended the method to the near infra-red, determining the epoxide value of individual resins and also following the loss of epoxide during the cure process. Studies in the near infra-red were also used to measure the hydroxyl concentration of the resins, a more difficult task than measurement of the epoxide concentration, since the inter- and intramolecular hydrogen bonds formed by the hydroxyl groups cause variations in intensity and wavelength in the spectrum. However, a measure of the hydroxyl concentration was obtained from a series of spectra for a specific resin at different temperatures, during cooling from 100 °C. The equilibrium positions for the hydrogen bonds are different at each temperature but one point common to all the spectra was observed (the isosbestic point) where the absorption was apparently independent of the concentration of the various hydrogen bonded structures. Using this point, at which Dannenberg regarded the absorption to be a function of the total hydroxyl concentration alone, the specific hydroxyl concentration of each resin was calculated.

2.4.3 RESIN CHEMICAL AND PHYSICAL CHARACTERISTICS

Characteristics often used by manufacturers to specify particular grades of resin are: (a) epoxide content, (b) viscosity, (c) softening point, and (d) colour. In addition, typical properties such as density, vapour pressure, flash point, refractive index, solubility characteristics, and hydroxyl content are frequently quoted.

The amount of epoxide groups present in a resin is usually expressed as the *weight per epoxide* (WPE) or *epoxide equivalent*, this being defined as the weight of the resin in grams which contains one gram equivalent of epoxide. For an unbranched diepoxide the WPE will therefore be half the number average molecular weight and, for a tetraepoxide, a quarter the molecular weight. The term 'epoxide content' is also used and expresses the amount of epoxide present as equivalents per kilogram of resin. This value is in fact the reciprocal of the WPE, multiplied by one thousand.

Epoxide equivalents can be determined by chemical or physical methods, and the use of infra-red spectroscopy for this has been described above. The chemical method of analysis usually depends upon the reaction of an hydrogen halide with the epoxide, yielding the halohydrin, thus:

$$\underset{\substack{| \quad |}}{-C\!-\!C-} \overset{O}{\diagup\!\!\diagdown} + \; HX \longrightarrow \underset{\substack{| \quad |}}{-\overset{OH}{C}\!-\!\overset{X}{C}-}$$

The consumption of the acid, a measure of the epoxide content, is obtained by using an excess of acid, with back-titration. There are a number of different methods using hydrogen fluoride, bromide, or iodide, and these are fully described by Gulinsky and Gruber,[50] Lee and Neville,[46] Knoll *et al.*,[51] and resin manufacturers' technical literature.[34] The success of the analysis depends upon the reaction of one molecule of the acid with one epoxide group; side reactions which consume epoxide groups without consumption of acid are undesirable.

In a direct method of estimation, the resin is dissolved in methyl ethyl ketone, glacial acetic acid added, and the titration carried out with 0.1N perchloric acid in glacial acetic acid, in the presence of cetyltrimethylammonium bromide. Crystal Violet is used as an indicator. Some typical values for the WPE of commercial resins are given in the Appendix to this Chapter.

The higher molecular weight glycidyl ethers of DPP (those with $n \geqslant 1$ in the general formula) have secondary hydroxyl groups spaced at regular intervals along the molecular chain. These groups can enter into reaction with fatty acids, as in esterification, or with isocyanates, and can influence the rate of polymerisation with certain curing agents. Knowledge of the hydroxyl concentration can therefore be valuable.

The hydroxyl content is usually expressed as *hydroxyl equivalent*, the weight of resin in grams containing one gram equivalent of hydroxyl (i.e., 16 g of oxygen present as hydroxyl), or as hydroxyl equivalent per kg, and is determined by near infra-red spectroscopy or chemical methods. The former has already been described; the chemical methods include reduction with lithium aluminium hydride, esterification with acid, or acetylation with acetyl chloride, a review of the subject being given by Bring and Kadlecek.[52] The

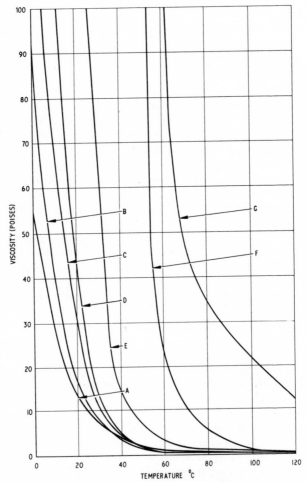

Fig. 2.13 Variation of viscosities of liquid epoxide resins with temperature. A = flexibilising resin based on dimerised fatty acid, B = liquid diglycidyl ether plus n-BGE reactive diluent, C = liquid diglycidyl ether plus glycidyl ester of a tertiary carboxylic acid as a reactive diluent, D = liquid diglycidyl ether plus dibutyl phthalate as plasticiser, E = unmodified diglycidyl ether of DPP, F = semi-solid diglycidyl ether, G = epoxidised novolak

first method depends upon the reaction:

$$R \cdot OH + LiAlH_4 \rightarrow LiAl(OR)_4 + 4H_2$$

The usual solvent is tetrahydrofuran, and the volume of hydrogen evolved is used to calculate the hydroxyl content of the system. Stenmark and Weiss[53] find values for the liquid diglycidyl ether resins of 0·091–0·094 equivalent per 100 g, and for the low molecular weight solid resins 0·259–0·264 equivalent per 100 g.

Other quantitative tests are carried out on the resins to determine α-glycol, phenolic hydroxyl, chlorohydrin, total chlorine, and easily saponifiable chlorine.

Liquid resin viscosities are usually measured by rotating cylinder or capillary instruments, and solid resins are mostly dissolved in a solvent at 40% concentration and the viscosity measured using bubble tubes or a capillary viscometer. Liquid resins do not show Newtonian behaviour, and their viscosity is dependent upon the rate of shear at which it is measured. The viscosity is also highly temperature-dependent, as illustrated in Fig. 2.13. Small changes in temperature can bring about large changes in viscosity, and this effect is very important when mixing and pouring resin systems.

The Durrans' mercury method of determining the melting point of resinous substances is usually employed with epoxides. This consists of melting a weighed sample of the resin in a standard test-tube and then cooling it. A weighed quantity of mercury is then placed on top of the solidified sample. The tube and contents are then heated again and the temperature at which the molten sample rises to the top of the mercury is recorded as the melting point.[53]

To measure the colour of resins, samples are dissolved in a solvent such as a glycol ether, and the colour of the solution visually compared with a standard colour-disc by a suitable colour comparator.

REFERENCES

1.   CASTAN, P., Brit. Pat. 518,057
2.   CASTAN, P., Brit. Pat. 579,698
3.   WERNER, E. C. G., and FARENHORST, E., U.S. Pat. 2,467,171
4.   GOPPEL, M., U.S. Pat. 2,801,227
5.   MOROSON, H. L., U.S. Pat. 2,921,049
6.   MOROSON, H. L., U.S. Pat. 2,943,095
7.   GREENLEE, S. O., U.S. Pat. 2,582,985
8.   GREENLEE, S. O., U.S. Pat. 2,615,007

9. GREENLEE, S. O., U.S. Pat. 2,615,008
10. GREENLEE, S. O., U.S. Pat. 2,694,694
11. GREENLEE, S. O., U.S. Pat. 2,698,315
12. PARTANSKY, A. M., and SCHRADER, P. G., *Am. chem. Soc. Symp.* (1956)
13. HOWE, B. R., IVINSON, M. G. and KARPFEN, F. M., *J. appl. Chem., Lond.* **7,** 118 (1957)
14. APPLEGATH, D. D., HELMREICH, R. F., and SWEENEY, G. A., *S.P.E.Jl.,* **15,** No. 1, 38 (1959)
15. Dow Chemical Co., U.S.A., Technical Literature
16. WHITEHOUSE, A. A. K., PRITCHETT, E. G. K. and BARNETT, G., *Phenolic Resins,* published for The Plastics Institute by Iliffe, London (1967)
17. BRADLEY, T. F., and NEWEY, H. A., U.S. Pat. 2,716,099
18. MOULT, R. H. and ST. CLAIR, W. E., *Am. chem. Soc. Symp.* (1957)
19. SCHWARZER, C. G., U.S. Pat. 2,806, 016
20. ZECH, J. D., U.S. Pat. 2,538,072
21. ANDERSON, C. C., *Ind. Engng. Chem.,* **60,** 8, 82 (1968)
22. MUELLER, A. C., U.S. Pat. 2,772,296
23. SHOKAL, E. C., U.S. Pat. 2,895,947
24. SCHRADE, P., Ger. Pat. 1,176,122
25. Farbenfabriken Bayer A.G., Brit. Pat. 772,830
26. Farbenfabriken Bayer A.G., Brit. Pat. 816,923
27. COOKE, H. G., U.S. Pat. 2,809,942
28. BUDNOWSKI, M., U.S. Pat. 3,337,509
29. BUDNOWSKI, M. and DOHR, M., U.S. Pat. 3,288,789
30. BUDNOWSKI, M., U.S. Pat. 3,300,490
31. BUDNOWSKI, M., Belg. Pat. 588,543
32. BUDNOWSKI, M., *Kunstsoffe,* **55,** 641 (1965)
33. EARHART, K. A. and MONTAGUE, L. G., *Ind. Engng. Chem.,* **49,** 1095 (1957)
34. Shell Chemical Co., Technical Literature
35. SPELL, H. L. and EDDY, R. D., *Am. chem. Soc. Symp.,* 148th meeting (1964)
36. WEATHERHEAD, R. G., *Analyst Lond.,* **91,** 445 (1966)
37. MOORE, J. C., *J. Polym. Sci.,* **A-2,** 835 (1964)
38. EDWARDS, G. D., *J. appl. Polym. Sci.,* **9,** 3845 (1965)
39. Waters Associates, Framingham, Mass., U.S.A., Technical Literature
40. BARTOSIEWICZ, R. L., *J. Paint Technology,* **39,** 504, 28 (1967)
41. MILES, B. H., *Am. chem. Soc. Symp.* (1964)
42. PATTERSON, W. A., *Analyt. Chem.,* **26,** 823 (1954)
43. BOMSTEIN, J., *Analyt. Chem.,* **30,** 544 (1958)
44. SHREVE, O. D., HEETHER, M. R., KNIGHT, H. B. and SWERN, D., *Analyt. Chem.,* **23,** 277 (1951)
45. LEE, H. L. and VINCENT, L., *Adhes. Age,* **4,** No. 9, 22 (1961)
46. LEE, H. L. and NEVILLE, K., *Handbook of Epoxy Resins,* McGraw-Hill, New York (1967)
47. KAGARISE, R. E. and WEINBERGER, L. A., U.S. Govt. Report PB 111438 (1954)
48. DANNENBERG, H. and HARP, W. R., *Analyt. Chem.,* **28,** 86 (1956)
49. DANNENBERG, H., *S.P.E. Trans.,* **3,** 78 (1963)
50. GULINSKY, E. and GRUBER, H., *Fette Seifen Anstrichmittel,* **59,** 1093 (1957)
51. KNOLL, D. W., NELSON, D. H. and KEHRES, P. W., *Am. chem. Soc. Symp.,* 134th meeting (1958)
52. BRING., A. and KADLECEK, F., *Plaste Kautsch.,* **5,** 2, 43 (1958)
53. STENMARK, G. A. and WEISS, F. T., *Analyt. Chem.,* **28,** 1784 (1956)

Appendix

SOME TYPICAL EPOXIDE RESINS AVAILABLE COMMERCIALLY IN THE U.K.*
(Source: Resin Manufacturer's Technical Literature)

BAKELITE LTD

Glycidyl ethers based on DPP

Grade	WPE	Viscosity Centipoises at 25°C	Comments
R.18774/1	180–200	8 500–12 500	Unmodified liquid resin
R.18774	180–205	3 000–5 500	Contains 1–4% of non-reactive diluent
R.19019	215–240	1 600–3 000	Contains plasticiser
R.19106	215–240	850–1 250	Contains 1–4% of non-reactive diluent

Grade	WPE	Softening point (ring and ball) 60–70°C	Comments
R.19102	420–450		Solid casting resin

BORDEN CHEMICAL CO. LTD

Glycidyl ether based on DPP

Grade	WPE	Viscosity Poises at 25°C	Comment
Epophen EL–5	200	75–175	Unmodified liquid resin

CIBA (ARL) LTD

Glycidyl ethers based on DPP

Grade	WPE	Viscosity Poises at 21°C	Comments
Araldite GY 250	192–196	225–275	Unmodified liquid resin
Araldite GY 251	230–238	20–40	Plasticised liquid resin
Araldite GY 252	185–196	7–15	Contains reactive diluent

Note: Both U.K. manufactured and imported resins are included in this list. Where a manufacturer producing epoxide resins in the U.K. has overseas affiliates and production facilities, only those resins mentioned in the U.K. sales literature are included. The list is not exhaustive but, it is hoped, typifies the range of products available and most widely used. Solutions of resins, cycloaliphatics, and some of the very specialised grades have, in general, been omitted.

Grade	WPE	Viscosity	Comments
Araldite GY 260	192–200	200–400	Unmodified liquid resin
Araldite GY 270	213–222	2 000–3 500	Unmodified liquid resin
Araldite GY 278	192–200	12–16	Contains reactive diluent
		Melting point °C	
Araldite 6100	*c.* 500	64–67	Solid resin
Araldite 6150	*c.* 650	76–82	Solid resin
Araldite 6200	*c.* 950	95–105	Solid resin
Araldite 6300	*c.* 1 900	127–133	Solid resin
Araldite 6400	*c.* 3 000	145–155	Solid resin

Epoxidised novolak

Grade	WPE	Viscosity
Araldite LY558	175–182	35–48

Glycidyl ethers based on brominated DPP

Grade	WPE		Comments
Araldite 8011	455–500	m.p. 70–80 °C	Solid resin
Araldite 8047	223–246	350–450	Semi solid resin
		centipoises at 70 °C	

DOW CHEMICAL CO. LTD
Glycidyl ethers based on DPP

Grade	WPE	Viscosity Centipoises at 25°C	Comments
DER 330	182–189	7 000–10 000	Unmodified liquid resin
DER 321	182–188	500–700	Contains reactive diluent
DER 331	186–192	11 000–14 000	General purpose unmodified liquid resin
DER 332	172–178	4 000–6 400	Highly purified diglycidyl ether of DPP
DER 332LC	170–175	3 000–6 400	Further purified DER 332 with low chlorine
DER 334	178–186	500–700	DER 331 + 10·5–11·5% n-BGE reactive diluent
DER 335	170–180	150–210	DER 331 + 19–21% n-BGE reactive diluent
DER 336	182–192	4 000–8 000	DER 331 + reactive diluent
DER 337	230–250	400–800 (as 70% solution in a solvent)	Semi-solid resin

Grade	WPE	Durrans m.p. (°C)	Comments
DER 660	425–475	65–74	Solid resin
DER 661	475–575	70–80	Solid resin
DER 662	575–700	80–90	Solid resin
DER 664	875–975	95–105	Solid resin
DER 667	1 600–2 000	113–123	Solid resin
DER 668	2 000–3 500	120–140	Solid resin
DER 669	3 500–5 500	135–155	Solid resin

Epoxidised Novolaks

Grade	WPE	Viscosity Centipoises at 52°C	Comments
DEN 431	172–179	1 400–2 000	
DEN 438	176–181	35 000–70 000	

Epoxidised Polyglycols

Grade	WPE	Viscosity Centipoises at 25°C	Comments
DER 732	305–335	55–100	Flexibilising resin
DER 736	175–205	30–60	Flexibilising resin

Glycidyl ethers based on brominated DPP

Grade	WPE	Durrans m.p. (°C)	Comments
DER 511	445–520	68–80	Solid resin containing 18–20% bromine
DER 542	350–400	51–61	Semi-solid resin containing 44–48% bromine

SHELL CHEMICAL CO. LTD
Glycidyl ethers based on DPP

Grade	WPE	Viscosity Poises at 25°C	Comments
Epikote 815	180–200	7–11	Contains reactive diluent
Epikote 816	185–205	12–18	Contains reactive diluent
Epikote 817	210–240	20–25	Contains plasticiser

Grade	WPE	Viscosity Poises at 25°C	Comments
Epikote 827	180–190	80–100	Unmodified liquid resin
Epikote 828	182–194	100–150	Unmodified liquid resin
Epikote 834	230–270	4–7 (as 70% solution)	Semi-solid resin
Epikote 1040	330–380	370–550 *(centipoises at 120°C)*	Low m.p. solid resin
		Durrans m.p. (°C)	
Epikote 1001	450–500	60–70	Solid resin
Epikote 1004	850–940	90–100	Solid resin
Epikote 1007	1 700–2 050	120–130	Solid resin
Epikote 1009	2 400–3 400	140–155	Solid resin

Epoxidised novolak

Grade	WPE	Viscosity Centipoises at 52°C	Comments
Epikote 154	176–181	35 000–70 000	

Flexibilising resins

Grade	WPE	Viscosity Poises at 25°C	Comments
Epikote 871	390–470	4–9	Aliphatic polyepoxide
Epikote 872	650–750	15–25 (as 75% wt. solution in xylene)	Aliphatic polyepoxide

Glycidyl ether based on brominated DPP

Grade	WPE	Viscosity Poises at 125°C	Comments
Epikote 1045–A–80	450–500	10–20	80% solution in acetone of resin containing 18–20% bromine

CHAPTER 3

Cure and Cure Mechanisms

3.1 INTRODUCTION

Before they can be used for most of their applications, epoxide resins need to be converted by means of cross-linking reactions into a three-dimensional infusible network held together by covalent bonds. In epoxide resin technology, this conversion from a liquid or friable brittle solid into a tough cross-linked polymer is called *curing* or *hardening*, and is achieved by the addition of a curing agent (hardener). The curing agents fall broadly into two types, *catalytic* and *polyfunctional*, and cross-linking between the resin molecules is achieved through the epoxide or hydroxyl groups of the resin. The catalytic curing agents serve as initiators for resin homopolymerisation, whereas the polyfunctional curing agents are used in near-stoichiometric amounts and function as reactants or comonomers in the polymerisation, leading mainly to the formation of a three-dimensional network composed of resin molecules cross-linked via the curing agents. In most of these curing reactions the actual mechanism is ionic in nature; both anionic and cationic polymerisations can occur, depending upon the curing agent concerned.

Cure normally occurs without the formation of by-products. The curing reactions are exothermic and the rates of reaction are increased by increases in temperature. Klute and Viehmann[1] measured the heats of polymerisation of liquid epoxide resins and found values of 22 kcal/mole epoxide when a tertiary amine was used as a curing agent and 25 kcal/mole epoxide when a primary amine was employed. Rates of reaction and activation energies have also been measured for certain polymerising systems, and are considered in Section 3.8.

The heat formed by the exothermic reaction can lead to a considerable rise in temperature of the system, known as the 'exotherm', and the actual levels reached depend not only upon the reactivity

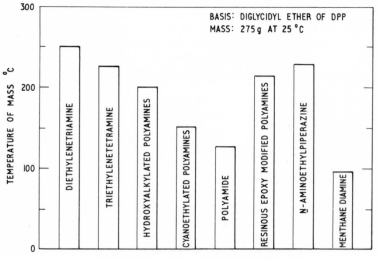

Fig. 3.1 Peak exotherms of epoxide-polyamine systems

of the resin and curing agent but also upon the temperature of the reactants and their surroundings, i.e. the rate at which polymerisation is occurring and the rate at which the heat evolved is being dissipated to the surroundings.

Some values for the temperature rise of a casting are shown in Fig. 3.1. It can be seen that quite high temperatures are reached (250 °C), and in practice this could lead to bubbling, damage to temperature-sensitive components, and in severe cases charring and complete degradation of the resin. It may even explode. Control of the exotherm is therefore an extremely important factor in the selection of a resin and curing agent. This is especially important when large castings are being made, where the heat evolved in the centre of the casting cannot readily escape owing to the low thermal conductivity of the resin.

Most curing agents will react at least partially with the resin at room temperature if given long enough. Thus the mixing of resin and curing agent is a point of no return; cure begins and proceeds at a rate determined by those factors which govern exotherm. The mixture of resin and curing agent has a finite *pot life* or *gel time,* beyond which the viscosity of the system has so increased as to render the materials unusable. In practice, pot lives can vary from a few seconds to two years or more, such is the wide range of possibilities in epoxide resin formulating. Many different methods have been developed to measure pot life, the most popular ones being based on the rotating cylinder viscometer principle. Values for pot life are also dependent upon the intended application of the system. A viscous mixture may still be pourable into a mould, but would not be suitable for glass fibre impregnation in a laminating process.

The epoxide ring is susceptible to attack from a wide variety of substances, ranging from those containing active hydrogen atoms such as phenols, alcohols, thiols, primary and secondary amines, carboxylic acids to cure catalysts such as tertiary amines, and Lewis bases and acids in general.

To obtain a three-dimensional meshed network on curing, the functionality of the resin and curing agent combination must accord with Kienle's principle that one component must have a functionality greater than two, and the other a functionality not less than two, their sum therefore being not less than five. In practice the functionality of groups can depend upon the particular set of reaction conditions being considered. Thus, although the epoxide group is regarded as having a functionality of two, only one of these is involved when it reacts with an amino hydrogen, the hydroxyl group not entering into any significant reaction under these conditions.

$$\overset{O}{\overset{\displaystyle\diagup\,\diagdown}{-CH-CH_2}} + NHRR' \longrightarrow \overset{OH}{\overset{|}{-CH}} \cdot CH_2 \cdot NRR'$$

However, when a Lewis base :X catalyses epoxide homopolymerisation, the functionality of each epoxide group can be taken to be 2 (Fig. 3.2).

Apart from the nature of the curing agent, the total environment of the epoxide group also has a great influence on the curing process. Steric factors can play an overriding part[2,3] in determining the ease with which a polymerisation can take place. In addition, the electronic influences exerted on the epoxide group are extremely

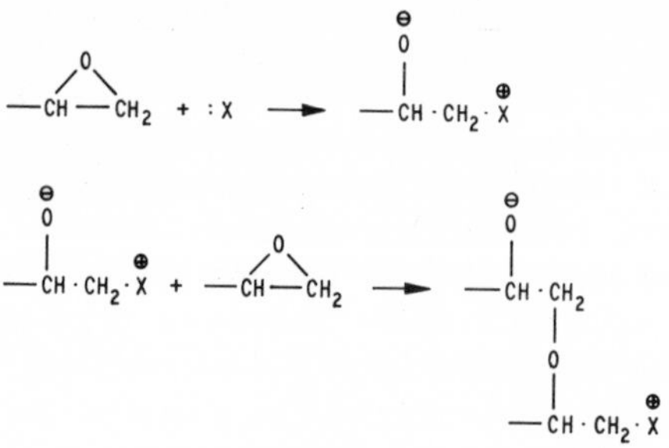

Fig. 3.2

important. Groups adjacent to the epoxide ring which withdraw electrons often enhance the reactivity of the epoxide to attacking nucleophilic reagents, reducing its reactivity towards electrophilic reagents.[2, 4, 5]

3.2 ALIPHATIC POLYAMINES AND POLYAMIDES

The reaction of an epoxide resin with this class of curing agent would be expected to follow the reaction sequences shown in Fig. 3.3. The reaction with a primary amine leads to the formation of a secondary hydroxyl group and a secondary amine. This secondary amine can then undergo reaction with another epoxide group to form a further secondary hydroxyl group and a tertiary amine. This tertiary amine could then act as a catalyst for epoxide homopolymerisation, but its effectiveness in this role is entirely dependent on its structure. The amine could be formed at a stage where it was effectively immobile. A further possible reaction is between the secondary hydroxyl and an epoxide leading to cross-linking via the formation of another secondary hydroxyl group which could undergo a similar reaction.

Shechter et al.[6] studied the reaction of the model epoxide compound phenyl glycidyl ether with the monofunctional primary and secondary amines n-butylamine and diethylamine respectively, and concluded that the alcohol-epoxide reaction (Fig. 3.3(c)) did not occur to a detectable extent, the exclusive reaction being that of amine with epoxide. This conclusion was borne out by Dannenberg,[7] who showed by near-infra-red spectroscopy that the sum of the hydroxyl and epoxide values remain fairly constant during cure, and also by the work of O'Neill and Cole.[8]

Whilst the amine-epoxide reaction does not require a catalyst, it is known that hydroxyl groups are able to accelerate it to varying degrees. Shechter investigated the role of hydroxyl groups and concluded that it was a catalytic one. Various hydroxyl-containing compounds accelerated the reaction rate to differing extents. Water and isopropanol, when added in molar quantities to equimolar amounts of phenyl glycidyl ether and diethylamine at 50 °C, had a pronounced effect on the rate of consumption of epoxide groups. Addition of a mole of phenol to the same reactants caused such a degree of acceleration that the reaction started at room temperature and rapidly got out of control.

In the same investigation it was found that addition of molar amounts of benzene or acetone had a retarding effect on the rate of

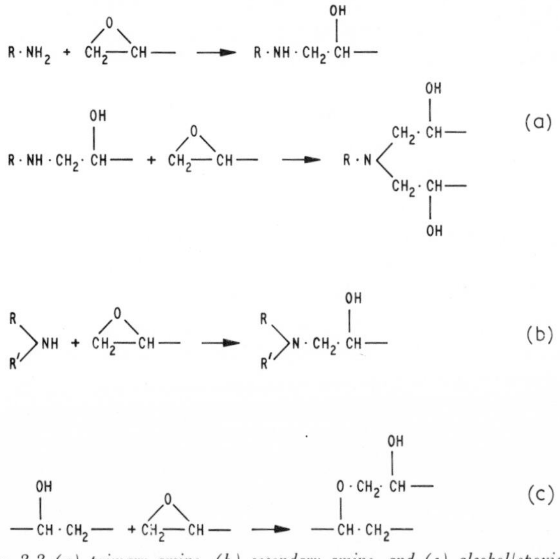

Fig. 3.3 (a) primary amine, (b) secondary amine, and (c) alcohol/epoxide

epoxide consumption. This led to the suggestion that the addition of these solvent molecules served to reduce the concentration of the reactants and hence depress the reaction rate, whilst the fact that acetone reduced the rate as much as did benzene demonstrated that the role of hydroxyl in assisting the epoxide-amine reaction was specific, and not one of increasing the polarity of the system. This specificity led Shechter[6] to propose a termolecular 'push-pull' mechanism for the reaction (Fig. 3.4).

The role of the hydroxyl is regarded as assisting in opening the epoxide ring by hydrogen bonding to the oxygen in the transition state. Gough and Smith[9] related the ability of the accelerator to

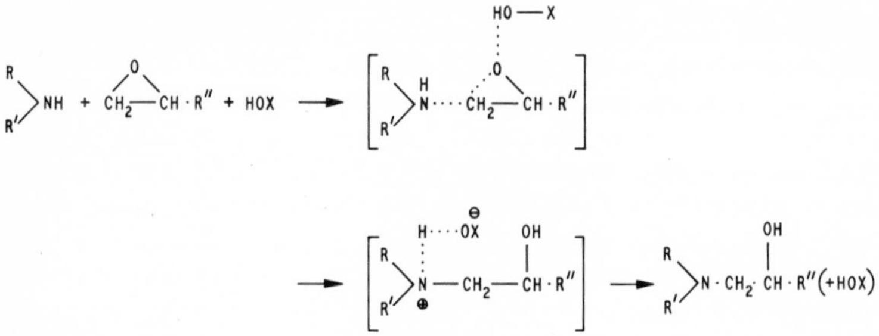

Fig. 3.4

act as a hydrogen donor to its effect on the rate of the epoxide-amine reaction. Groupings which are hydrogen donors and should therefore act as accelerators for the reaction are:

$$-OH \qquad -CO_2H \qquad -SO_3H \qquad -CONH_2 \qquad -SO_2NH_2$$

Conversely, cure retarders (hydrogen acceptors) should include:

$$-OR \qquad -CO_2R \qquad -SO_3R \qquad -CO- \qquad -CN \qquad -NO_2$$

Smith[3] also considered that the function of the solvent in the reaction might be (*a*) to weaken the C–O bond by hydrogen bonding to the epoxide oxygen, (*b*) to affect the ease of charge-separation in the transition state, and (*c*) to take part in the proton exchanges with amine and product.

The first stage of the amine-epoxide reaction produces a secondary alcohol, and this too may form a hydrogen bond with the epoxide group. Hence, if the concentration of solvent is small enough, auto-catalysis of the reaction may be observed, and Smith proposed the mechanism shown in Fig. 3.5 to explain the experimental evidence. The rapid formation of a hydrogen bond between HX and the epoxide oxygen is followed by the opening of the epoxide ring via the termolecular transition state, this being the rate-determining step. Finally the sequence is completed by fast proton displacements. The S_N2 mechanism proposed, which involves simultaneous attack of the nucleophilic reagent (polyamine) at the methylene group of the epoxide and displacement of the ring oxygen, is supported by the second order kinetics observed in most epoxide-polyamine systems.[4, 5]

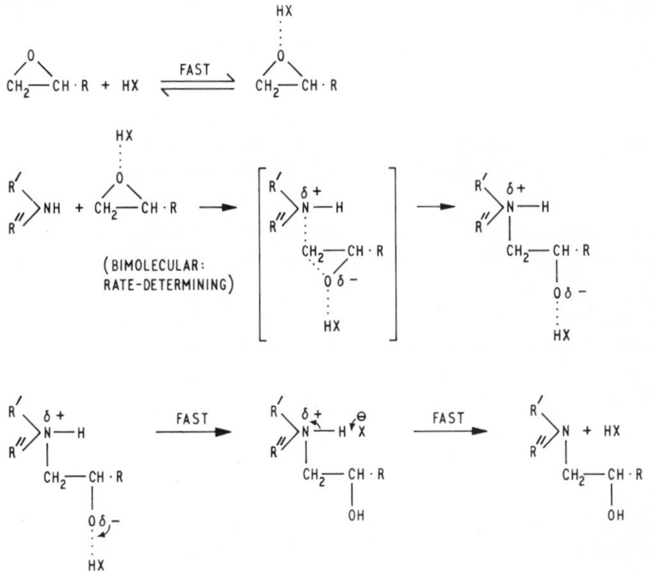

Fig. 3.5

The reactivity of an amine towards an epoxide group is also affected by steric factors which might tend to shield the functional groups, sterically hindered amines exhibiting slower rates of reaction than unhindered ones.[6, 10]

Shechter *et al.*[6] also briefly examined the reaction of two moles of phenyl glycidyl ether with one mole of butylamine at 50 °C, to determine whether the conversion of primary amine to secondary was faster than secondary to tertiary. If this were the case, and there was a significant difference between the rates, the plots of the concentration of the various amine types against time would show a stepwise progression. This was not found to be so. Almost immediately some tertiary and secondary amines were formed in the reaction mixture, indicating that the conversion reactions proceed more or less at random, there being no great selectivity in any of the reactions.

The polyamides most widely used with epoxide resins are derived from the reaction of polymeric fatty acids with aliphatic polyamines. They contain a number of secondary amino groups and terminal primary amino groups in addition to tertiary amino, amide, and imidazole groups. Cure probably proceeds via the first two groups by the mechanisms already discussed.

3.3 AROMATIC PRIMARY AND SECONDARY POLYAMINES

In general, the chemistry of cure with aromatic amines is similar to that of the aliphatic amines, though the lower basicity of the former has some important consequences. A typical aromatic amine curing agent, *m*-phenylenediamine (MPD), might undergo reaction with two molecules of a glycidyl ether resin initially as shown in Fig. 3.6.

Further reaction between resin molecules and the hydrogen atoms of the partly reacted amine would follow, accompanied by the terminal epoxide groups at (a) and (b) reacting with other molecules of MPD. Hence a fairly rigid three-dimensional network would

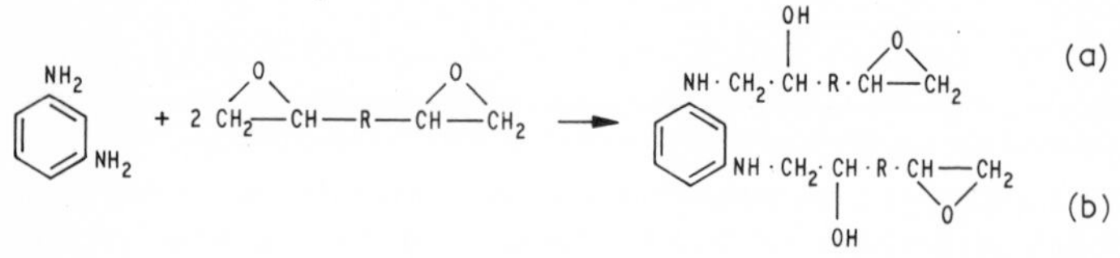

Fig. 3.6

quickly be built up. If the reaction is carried out at room temperature, this leads to a soluble, fusible, partly reacted resin with about 30% of available epoxide groups consumed, further reaction being prevented by the lack of mobility of the molecules. The resin would then be regarded as having reached the B stage of cure, and have a usable pot life of about 4 weeks. Further cross-linking can then be achieved by curing at elevated temperatures. Because of this limitation of the degree of polymerisation, aromatic amines do not show high exotherms at room temperature. If used in a large mass however, or cured above the melting point of the initial polycondensation product, exotherms are developed.

Aliphatic amine-epoxide resin systems do not form such insoluble high melting initial reaction products, and in these cases cure can proceed at room temperature to about 60% consumption of epoxide groups before the degree of polymerisation prevents further conversion. The formation of B-stage systems by means of close control of the reaction between aromatic amines and epoxide resins is used in a number of practical applications.

Aromatic amines react faster with cycloaliphatic epoxides than do aliphatic amines. This is probably due to the greater acidity of the aromatic compounds. Although, in general, aromatic amines require cure at elevated temperatures, they can be accelerated to approach the reaction rates of some aliphatic amines and in fact, when used in certain solvents, can polymerise glycidyl ether epoxides at room temperature.

3.4 POLYMERCAPTANS

Analogous to the hydroxyl group, the thiol or mercapto group (—SH), reacts with an epoxide group forming an hydroxy sulphide:

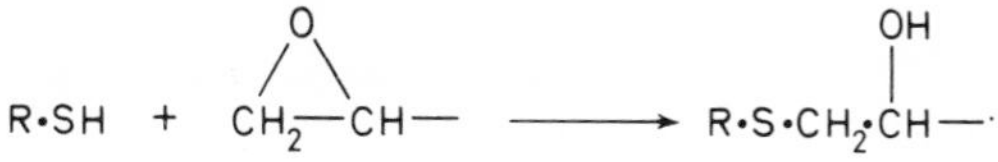

$$R \cdot SH \ + \ CH_2\!\!-\!\!CH\!\!- \ \longrightarrow \ R \cdot S \cdot CH_2 \cdot CH\!\!-$$

Hence polymercaptans can be used as cross-linking agents for epoxide resins, provided that the functionality of the mercaptan is equal to or greater than three or that cross-linking through the hydroxyl group on the resin can also be effected. Shokal[11] has used this reaction to prepare linear sulphides of high molecular weight, the thiol groups formed when the epoxide rings of a resin are opened with hydrogen sulphide reacting with further resin molecules (Fig. 3.7).

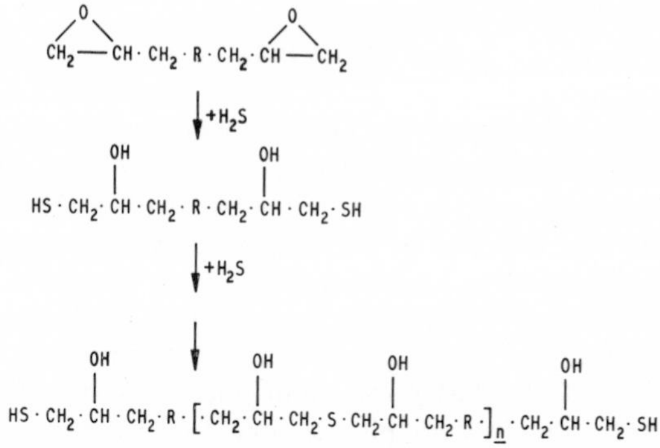

Fig. 3.7

Contrary to earlier belief, the epoxide-mercaptan reaction is several times faster than the epoxide-amine reaction, especially at low temperatures. It is also accelerated by a range of primary and secondary amines such as triethylenetetramine, piperidine, and

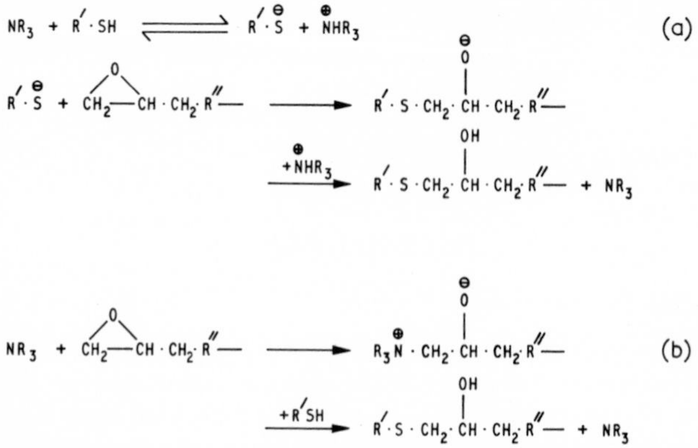

Fig. 3.8

m-phenylenediamine. This catalytic effect may operate through the mercaptide ion[12] (route (a) in Fig. 3.8) or the amine could open the epoxide ring (route (b)).

Steric effects play an important part in the efficiency of the amine catalysts in the epoxy-polymercaptan reaction. Systems where the amine groups of the catalyst are hindered show lower rates of epoxide consumption than those for less hindered systems. Increasing the basic strength of the amine also increases the reaction rate.

3.5 POLYPHENOLS

The reaction between phenolic hydroxyl and epoxide, when catalysed by caustic soda, is used in the synthesis of the higher molecular weight epoxide resins. There are two possible reactions, the phenol-epoxide followed by the alcohol-epoxide reaction (Fig. 3.9). Shechter and Wynstra,[2] working with model compounds, found that in the absence of catalysts no reaction occurred between epoxide and phenol at 100 °C. At 200 °C reaction commenced and epoxide was consumed at a faster rate than phenol. About 60% of the reaction was epoxide with phenol and 40% epoxide with alcohol. Since alcohol was absent at the beginning of the reaction and only appeared when phenol reacted with epoxide it was concluded that the phenol preferred to catalyse the alcohol-epoxide reaction rather than react itself.

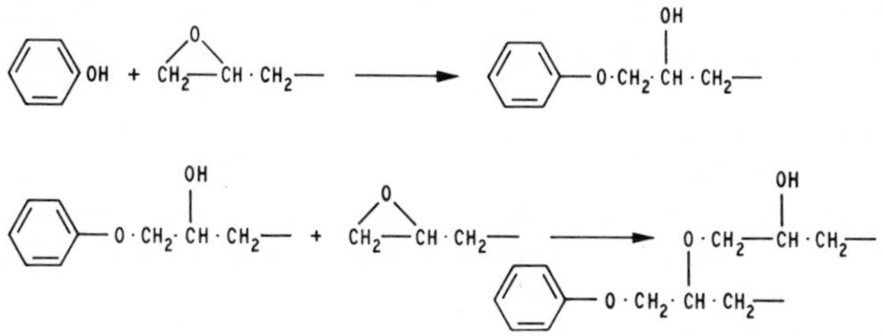

Fig. 3.9

The base-catalysed reaction, however, proceeded at 100 °C almost exclusively via the phenol-epoxide route, to the virtual exclusion of the alcohol-epoxide reaction. Shechter and Wynstra proposed the mechanism shown in Fig. 3.10 to fit the experimental facts.

In the same study, the role of tertiary amines as catalysts for the reactions was also examined and it was observed that benzyl-dimethylamine was a more effective catalyst than potassium

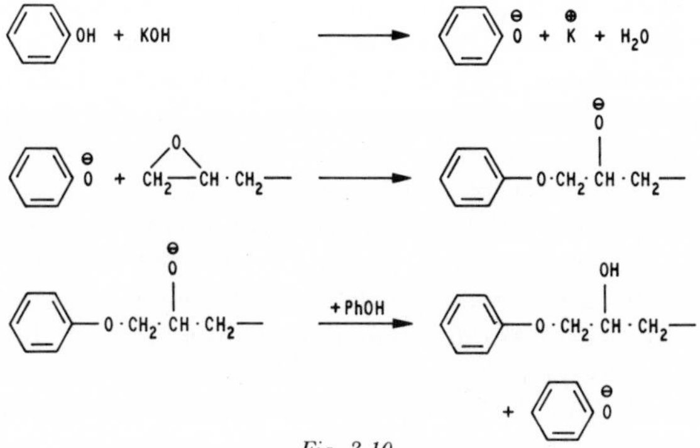

Fig. 3.10

hydroxide, and the quaternary compound benzyltrimethyl-ammonium hydroxide was even more powerful. In each case, the epoxide-phenolic reaction was essentially the only reaction occurring and first order kinetics were observed.

3.6 CATALYTIC CURE

3.6.1 LEWIS BASES

These substances contain an unshared pair of electrons in an outer orbital which are available for bond formation. They are therefore nucleophilic in character, seeking areas of low electron density. As catalytic curing agents they are used in small amounts, and primarily achieve homopolymerisation of the resin. In epoxide resin technology the most important compounds of this type are tertiary amines, although they have only found wide use in conjunction with other curing agents, usually primary or secondary amines, or acid anhydrides. However, imidazoline derivatives give indications of becoming an important group of curing agents in their own right. Frequently curing agents will contain both tertiary and primary amino groups in the molecule, the primary amine moiety also becoming a tertiary amine once its hydrogen atoms are consumed by the cross-linking reactions.

Mechanisms for the polymerisation of epoxide resins by tertiary amines have been proposed by Narracott,[13] Newey,[14] and Shechter

Fig. 3.11

and Wynstra.[2] Narracott proposed that the reaction was initiated by the tertiary amine attacking the epoxide ring carbon atom causing the formation of an alkoxide ion (Fig. 3.11).

From their observations on the effect of alcohol concentration on reaction rate, Shechter and Wynstra postulated that the alkoxide ion underwent a reaction with the alcohol present to form a further alkoxide ion (Fig. 3.12) and that this step was the rate-determining one. The ion formed (Fig. 3.11) was thought not to be an effective catalyst because of the close proximity of its centres of charge.

Fig. 3.12

Polymerisation can then be continued via the alkoxide ion (Fig. 3.13).

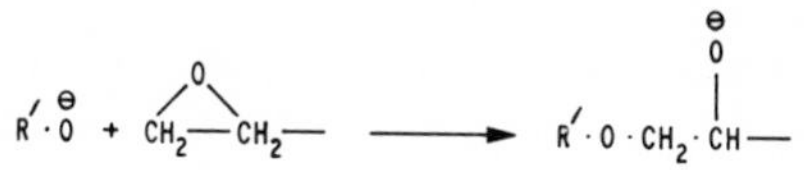

Fig. 3.13

Newey obtained epoxide polymerisation in the virtual absence of hydroxyl groups, and has proposed that the reaction in Fig. 3.11 is indeed the initiating step to be followed by the propagating step (Fig. 3.14).

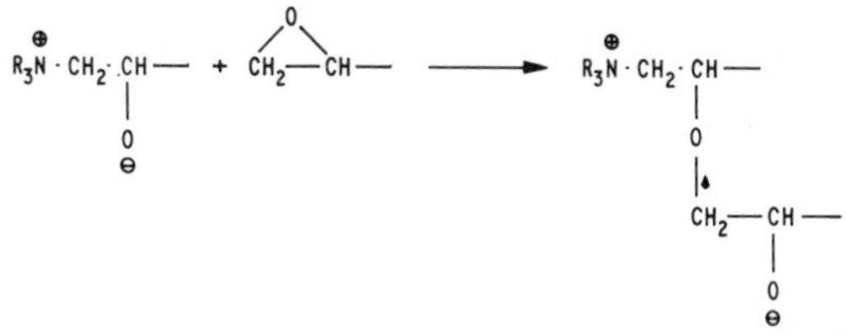

Fig. 3.14

As with primary and secondary amines, the rate of cure of epoxide resins with tertiary amines depends to a large degree upon the extent to which the nitrogen atom is sterically blocked. Newey found that steric factors played a more important part in determining the reaction rate than did the basicity of the amines, and that alcohols accelerated the curing reaction.

3.6.2 INORGANIC BASES

Alkali metal hydroxides such as sodium or potassium hydroxide also catalyse the polymerisation of epoxide resins (Fig. 3.15), the reaction rate being effectively independent of the concentration of any alcoholic hydroxyl present.[2]

When phenol is present in the reaction mixture the alkali serves as a very specific catalyst for the epoxide-phenoxide reaction, all other reactions being virtually excluded until all phenolic hydroxyl

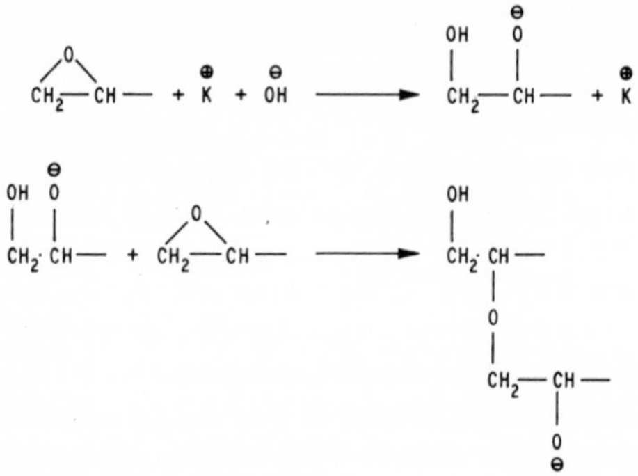

Fig. 3.15

groups are consumed. This is one reason why sodium hydroxide has almost always been used as the catalyst in the manufacture of glycidyl ether epoxide resins based on DPP and ECH.

3.6.3 LEWIS ACID CATALYSTS

Lewis acids contain empty electron orbitals in the outer shell of an atom which can be used in forming a bond by another atom donating both electrons to the bond; they therefore seek areas of high electron density and are electrophilic in character (electron acceptors).

The only substance of this type to have found major use in curing epoxide resins is boron trifluoride, which is a highly corrosive gas capable of polymerising epoxide resins extremely rapidly—perhaps in a few seconds at room temperature. To obtain a material suitable for handling in practical situations, the complexes formed between boron trifluoride and amines or ethers are used. The most popular is that with ethylamine, a crystalline substance of indefinite shelf

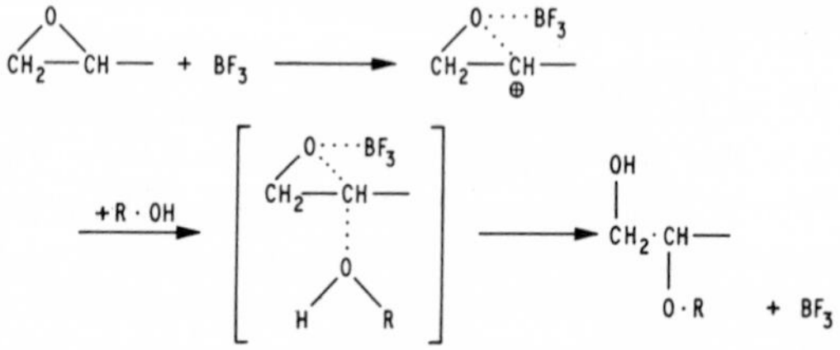

Fig. 3.16

life which decomposes into its active components when heated to 80–100 °C.

Mechanisms for polymerising epoxides with BF_3 in the presence of hydroxyl groups have been proposed by Landua,[15] Arnold,[16] and others. All assume initial attack by the BF_3 on the epoxide oxygen with the formation of a carbonium ion. This is followed by inter-action of hydroxyl with the ion, forming an alkoxide, with re-generation of the BF_3 (Fig. 3.16). However, Harris and Temin[33] have proposed an alternative theory for BF_3 cures which does not postulate the dissociation of the BF_3 complex.

In the usual situation, when the BF_3 is complexed with an amine, Arnold has proposed the sequence of reactions shown in Fig. 3.17. This requires loss of a proton by thermal dissociation, followed by the formation of a carbonium ion. (The importance of temperature

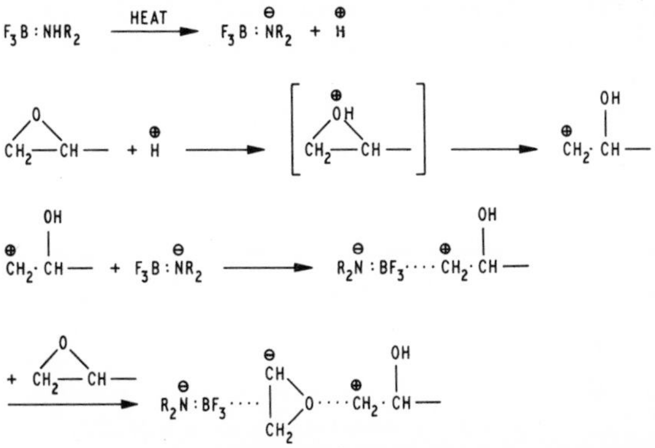

Fig. 3.17

in the application of such 'latent catalysts' is discussed in Section 4.6.2.) The ion is stabilised by the BF_3 complex and then undergoes reaction with an epoxide group, which is the propagating step.

3.7 POLYBASIC ACIDS, ACID ANHYDRIDES AND DERIVATIVES

Whilst polybasic carboxylic acids have not found wide use as curing agents, dicarboxylic acid anhydrides are next in importance to the amines. The reactions between epoxides and the acids can be expected to be of four types (Fig. 3.18). Reaction (a) leads to the formation of an hydroxy ester which, via reaction (b), can combine with a second molecule of acid forming a diester and water. The hydroxy ester can also serve to initiate polymerisation via the epoxide-hydroxyl reaction (c). Finally, the epoxide ring can undergo hydrolytic opening with water, forming the diol (d). It is probably the water sensitivity of the hydroxy ester (a) and its readiness to undergo hydrolysis, coupled with the fact that water is formed in reaction (b), that has limited the wider use of acids as curing agents.

Working with a low molecular weight diglycidyl ether liquid resin and caprylic acid in the absence of base catalysis, Shechter and Wynstra[2] established that reactions (a), (b), and (c) occurred in the ratio 2:1:1. When a PGE-caprylic acid system was examined in the presence of base catalysts reactions (e), (f), and (g) (Fig. 3.18)

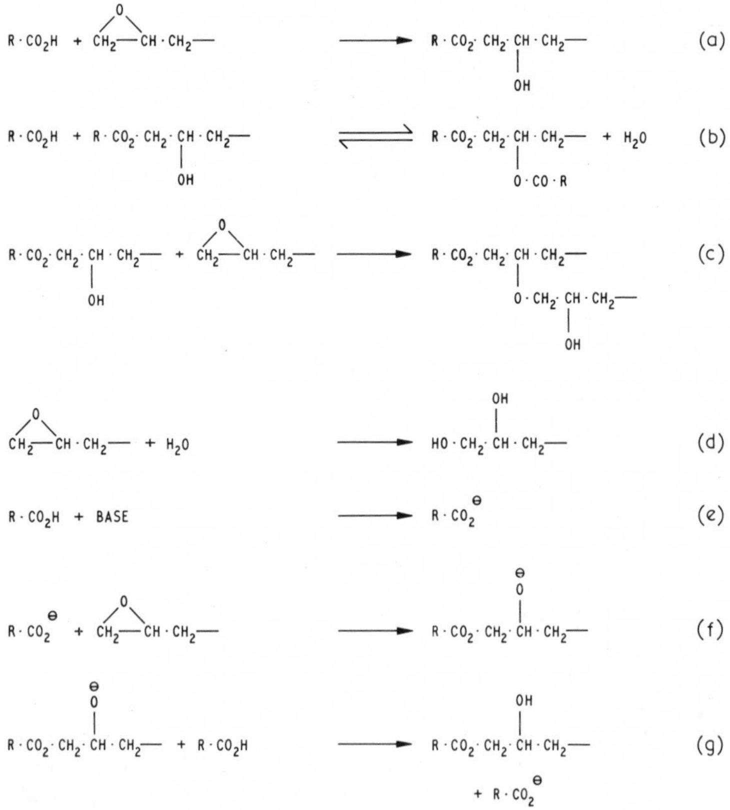

$$R \cdot CO_2H + CH_2\!-\!CH \cdot CH_2\!-\!\! \qquad \longrightarrow \qquad R \cdot CO_2 \cdot CH_2 \cdot CH \cdot CH_2\!-\!\! \qquad (a)$$
$$\underset{OH}{|}$$

$$R \cdot CO_2H + R \cdot CO_2 \cdot CH_2 \cdot \underset{\underset{OH}{|}}{CH} \cdot CH_2\!-\!\! \qquad \rightleftharpoons \qquad R \cdot CO_2 \cdot CH_2 \cdot \underset{\underset{O \cdot CO \cdot R}{|}}{CH} \cdot CH_2\!-\!\! + H_2O \qquad (b)$$

$$R \cdot CO_2 \cdot CH_2 \cdot \underset{\underset{OH}{|}}{CH} \cdot CH_2\!-\!\! + CH_2\!-\!CH \cdot CH_2\!-\!\! \qquad \longrightarrow \qquad R \cdot CO_2 \cdot CH_2 \cdot \underset{\underset{O \cdot CH_2 \cdot \underset{\underset{OH}{|}}{CH} \cdot CH_2\!-}{|}}{CH} \cdot CH_2\!-\!\! \qquad (c)$$

$$CH_2\!-\!CH \cdot CH_2\!-\!\! + H_2O \qquad \longrightarrow \qquad HO \cdot CH_2 \cdot \underset{\underset{OH}{|}}{CH} \cdot CH_2\!-\!\! \qquad (d)$$

$$R \cdot CO_2H + BASE \qquad \longrightarrow \qquad R \cdot CO_2^{\ominus} \qquad (e)$$

$$R \cdot CO_2^{\ominus} + CH_2\!-\!CH \cdot CH_2\!-\!\! \qquad \longrightarrow \qquad R \cdot CO_2 \cdot CH_2 \cdot \underset{\underset{O^{\ominus}}{|}}{CH} \cdot CH_2\!-\!\! \qquad (f)$$

$$R \cdot CO_2 \cdot CH_2 \cdot \underset{\underset{O^{\ominus}}{|}}{CH} \cdot CH_2\!-\!\! + R \cdot CO_2H \qquad \longrightarrow \qquad R \cdot CO_2 \cdot CH_2 \cdot \underset{\underset{OH}{|}}{CH} \cdot CH_2\!-\!\! \qquad (g)$$
$$+ R \cdot CO_2^{\ominus}$$

Fig. 3.18

occurred, the overall sequence became more specific, and the relative priorities of each depended upon the stoichiometry. Thus in general, with a ratio of epoxide to acid of 1:1, reactions (a) and (b) would occur, with formation of the hydroxy ester. This would virtually exclude any further reactions of the hydroxy ester, such as reaction (c) (Fig. 3.18), the formation of polyethers leading to resin homopolymerisation. Where an excess of epoxide groups occurs, reactions (a) and (b) would proceed until all acid was consumed. Thereafter, the faster epoxide-alcohol reaction (c) would commence.

Acid anhydrides react with epoxide groups in a similar way to the carboxylic acids. In the absence of catalysts, equimolecular amounts of epoxide and anhydride interact only sluggishly at 200 °C, the epoxide groups disappearing twice as fast as anhydride groups.[2] Fisch et al.[17, 18] propose the following series of reactions (Fig. 3.19) for this system. The anhydride ring is opened by a hydroxyl group (e.g. a secondary hydroxyl found on the higher molecular weight resins) forming a half-ester (Fig. 3.19(a)). This half-ester then

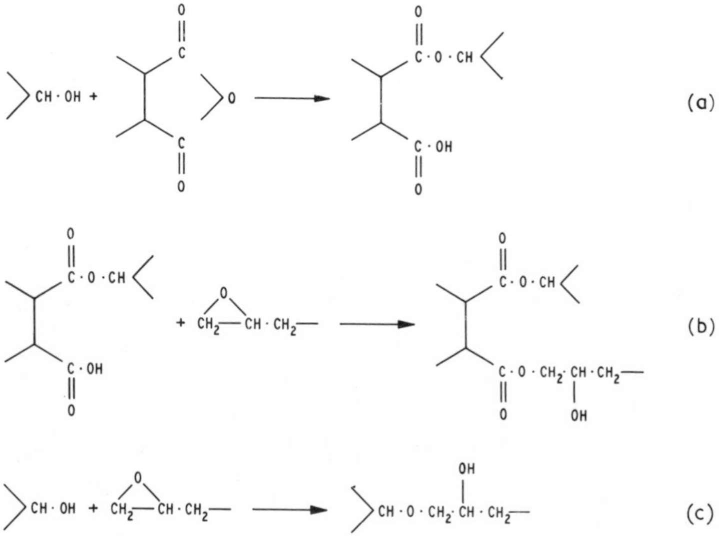

Fig. 3.19

reacts through its acid group with the epoxide ring forming a hydroxy diester (Fig. 3.19(b)). The hydroxyl group of the diester can undergo reaction with anhydride to form another carboxyl group, eventually yielding exclusively diester groups. However, in studying the phthalic anhydride/epoxide resin system, Fisch and Hofmann[17] found that the epoxide groups were consumed at a greater rate than could be accounted for by reaction (Fig. 3.19(b)) alone, and they therefore postulated, as a possible cause, an epoxide-hydroxyl etherification reaction (Fig. 3.19(c)) under the catalytic influence of carboxyl or anhydride groups. Dearborn[19] proposed a more general reaction for the removal of epoxide groups, and Shechter and Wynstra[2] supported these findings:

$$R \cdot H + \overset{\displaystyle O}{\overset{\displaystyle \diagup\!\!\!\diagdown}{CH_2\!-\!CH-}} \longrightarrow R \cdot CH_2 \cdot \overset{\displaystyle OH}{\underset{}{CH-}}$$

Tertiary amines are also able to open anhydride rings and hence act as catalysts for anhydride/epoxide reactions. For this system Fischer[20] proposed the sequence of reactions shown in Fig. 3.20. The base opens the anhydride ring, forming the carboxylate ion, which in turn reacts with an epoxide group yielding the alkoxide ester (Fig. 3.20(b)). This anion undergoes reaction with an anhydride, resulting in the formation of the ester (Fig. 3.20(c)). Continuation of the following alternating sequence leads to the formation of a polyester:

carboxylate ion + epoxide   →   alkoxide ester
alkoxide ester + anhydride   →   carboxylate ion

Tanaki and Kakiuchi[21] concluded that with base-catalysed systems

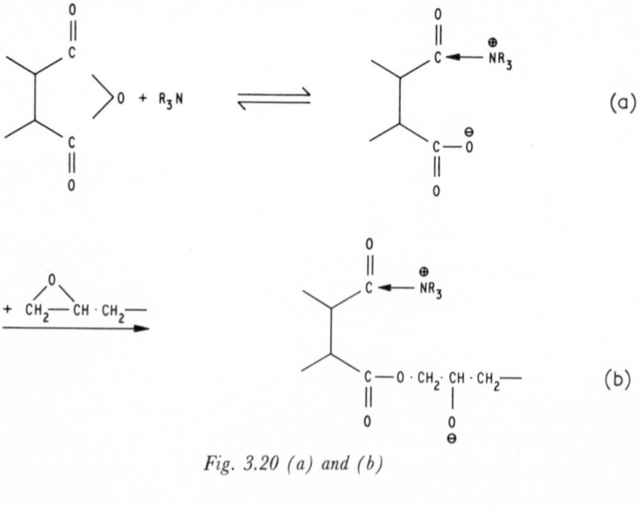

Fig. 3.20 (a) and (b)

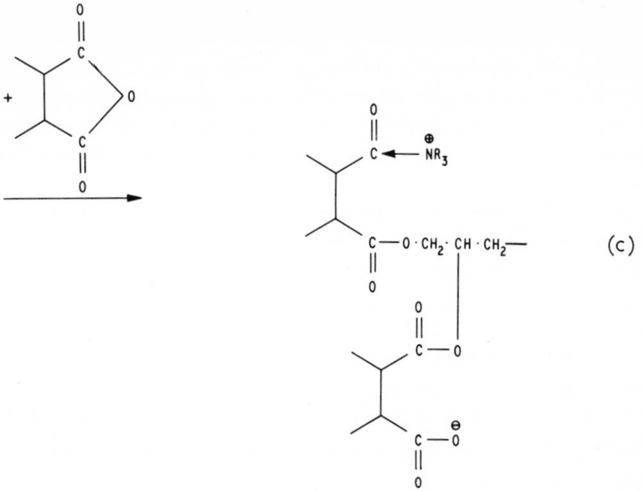

Fig. 3.20 (c)

etherification did not occur between 70°C and 140°C, and also found that the curing reaction rate was proportional to (i) the initial concentration of amine catalysts and (ii) the initial amount of hydroxyl in the resin.

Boron trifluoride and other Lewis acids are also accelerators for

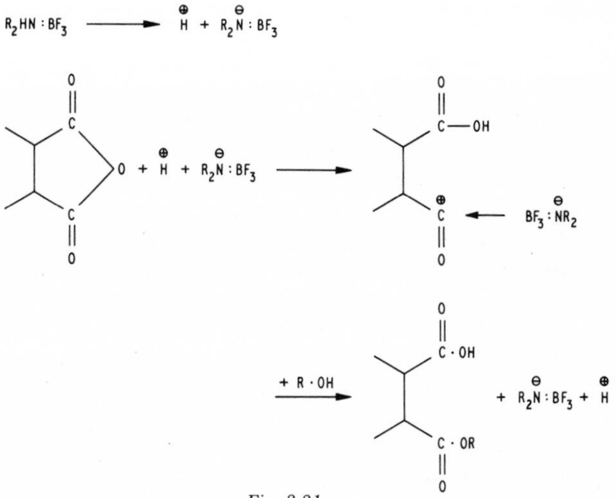

Fig. 3.21

the epoxide/anhydride reaction. The anhydride ring may thus be opened with these substances, as shown in Fig. 3.21.

In summary, the anhydride ring must be opened before reaction can occur with an epoxide group. This ring opening can be achieved by (1) hydroxyl groups such as those present in the higher molecular weight epoxide resins; (2) tertiary amines and other Lewis bases; (3) boron trifluoride complexes and other Lewis acids.

In case (1), when no catalyst is added, both esterification and etherification occur (Fig. 3.18, reactions (a) and (c)). When base catalysts are added (case (2)), esterification is favoured, the structure and concentration of the catalyst having an effect on the rate and specificity of the reaction. When acid catalysts are used (case (3)) etherification is promoted.

Etherification is undesirable since it results in unreactive anhydrides and/or acid monoesters being left in the cross-linked resin. For the desirable polyesterification to occur the optimum anhydride: epoxide stoichiometry is thought to be 1:1. However, this is a very complex reaction mechanism; and, depending upon the structure of the resin and anhydride, the anhydride:epoxide ratio used can vary from 0·4 to 1·2. Arnold[16] proposed, from collected experimental evidence, that for optimum properties the following anhydride:epoxide ratios were needed: 1:1 (for tertiary amine catalysts), 0·55:1 (for acid catalysts), 0·85:1 (no catalyst).

In practice, optimum ratios are best worked out experimentally for each specific system.

3.8 METHODS OF FOLLOWING THE POLYMERISATION PROCESS

There are numerous methods for following the process of cure in thermosetting materials, many of which have been used to investigate epoxide resin polymerisation. They include measurements of volume resistivity and other electrical properties, refractive index, infra-red absorption, dynamic mechanical properties and gel times.

Delmonte[22] measured volume resistivity, dielectric constant, and dissipation (loss) factor at different frequencies, for various combinations of resins and curing agents which were undergoing polymerisation. The most useful property for following the course of the cure reaction was the dissipation factor, which showed a sharp drop at the commencement of cure, increased to a peak when gelation set in, and then fell as cure continued. The peak value of the dissipation factor was found to be characteristic of the resin with each curing agent used. The polymerisation of epoxide resins was also followed by Warfield and Petree[23] by using the change in volume resistivity with time under isothermal conditions. Curves of resistivity versus time at different temperatures allowed the activation energy for the overall polymerisation process to be determined via Arrhenius plots. Cure was assumed to be complete when the resistivity became constant with time, and the specific polymerisation rate assumed to be proportional to the maximum rate of change of $\log_{10}$ (resistivity). Values of the activation energy determined in this way were 17 kcal/mole for a catalytic curing system and 14 kcal/mole for an aliphatic polyamine curing system. The glass transition point for the system was also determined, assuming that the transition occurred at the point where there was a significant change in the slope of the resistivity curve over a short temperature interval. Resistivity/time measurements were also used by Miller[24] to determine energies of activation, values of 14–17 kcal/ mole were obtained.

Dannenberg[25] reported that the change in refractive index of a polymerising system could also be used as a method of determining cure rates and hence the reactivity of the components and the extent of conversion.

The curing process has also been followed by Lewis and Gilham[26] using a braided nylon filament impregnated with the polymer solution, the filament serving as the supporting member for the reacting polymer system which was undergoing free torsional vibrations. The change in rigidity of the polymer was followed nondestructively throughout the cure by using the frequency of vibration to calculate the apparent rigidity modulus.

The polymerisation of the diglycidyl ether resins with aliphatic and aromatic polyamines as curing agents was studied by Jenkins and Karre[27] using both infra-red absorption and dynamic mechanical properties measured by a vibrating reed technique. Kinetic data for the polymerisation were obtained by the continuous *in situ* monitoring of the mechanical and spectroscopic properties at different temperatures. Hence activation energies of 11·4 kcal/mole for the aliphatic system and 10·4–11·0 kcal/mole for the aromatic system were determined. These values were higher than those obtained by Lewis[28] for the amine/epoxide reaction (8·5 ± 0·7 kcal/mole), but more in line with the results of Gough and Smith[29] who measured the activation energy for the amine-cured system and found it to be 12–17 kcal/mole, by using gel time measurements under isothermal conditions. Similar measurements, but at different temperatures, have been used for the same systems by Kakurai and Nogushi[30] who found activation energies of 12–13 kcal/mole. The progress of cure in paint films has also been studied using the infra-red spectra of reflected light and the change in elastic modulus as determined by a vibrating reed technique.[31]

Edwards[32] studied the exotherm characteristics of a large number of epoxide resin systems and used his results to predict the curing behaviour of a casting system under a variety of conditions. His method was to follow accurately the temperature rise of a reacting system of resin-curing agent, plus any other modifying materials, under conditions of known and preferably low heat loss. The exotherm curve (temperature/time curve) was used to determine graphically the rate of change of temperature at 5 degC intervals along the curve. After correction for heat losses, these values were a measure of the rate of heat output of the reacting system. By plotting the logarithm of the corrected rate of temperature increase against temperature, an 'effective reactivity' curve was obtained which was either a straight line or a smooth curve. The effective reactivity line, together with the cooling rate curve for the system, can then be used to determine whether the system will 'exotherm' or cure at a more or less constant temperature.

The kinetic data obtained from these and other studies show most curing reactions between an epoxide resin and a polyamine to have second order kinetics. This supports the S_N2 mechanism of attack by the amine (nucleophile) at the methylene group of the epoxide ring, accompanied by simultaneous displacement of the ring oxygen.[5] The kinetic measurements made on the reaction between the resins and anhydrides have not been so clear cut but the evidence of Tanaki and Kakiuchi[21] and Arnold[16] indicates the reaction rates to be proportional to the concentration of the catalytic species present, i.e. first order kinetics.

The activation energies of the amine curing reactions appear to be from 8 to 14 kcal/mole, depending upon the nature of the amine and the method of measurement. For anhydride curing agent, Tanaki and Kakiuchi[21] reported an overall activation energy of about 14 kcal/mole for the curing reaction between an epoxide resin containing hydroxyl groups and HPA in the presence of a catalyst.

REFERENCES

1. KLUTE, C. H. and VIEHMANN, W., *J. appl. Polym. Sci.*, **5,** 86 (1961)
2. SHECHTER, L. and WYNSTRA, J., *Ind. Engng. Chem.*, **48,** 86 (1956)
3. SMITH, I. T., *Polymer*, **2,** 95 (1961)
4. PARKER, R. E. and ISAACS, N. S., *Chem. Rev.*, **59,** 737 (1959)
5. CHAPMAN, N. B., PARKER, R. E. and ISAACS, N. S., *J. chem. Soc.*, **2,** 1925 (1959)
6. SHECHTER, L., WYNSTRA, J. and KURJY, R. P., *Ind. Engng. Chem.*, **48,** 94 (1956)
7. DANNENBERG, H., *S.P.E. Trans.*, **3,** 78 (1963)
8. O'NEILL, L. A. and COLE, C. P., *J. appl. Chem., Lond.*, **6,** 356 (1956)
9. GOUGH, L. J. and SMITH, I. T., *J. Oil Colour Chem. Ass.*, **43,** 409 (1960)
10. INGBERMAN, A. K. and WALTON, R. K., *J. Polym. Sci.*, **28,** 468 (1958)
11. SHOKAL, E. C., U.S. Pat. 2,633,458
12. MIKA, T. F., Private Communication
13. NARRACOTT, E. S., *Br. Plast.*, **26,** 120 (1953)
14. NEWEY, H. A., Gordon Research Conf. on Polymers, New London, U.S.A. (1955)
15. LANDUA, A. J., *Am. chem. Soc. Symp.* (1964)
16. ARNOLD, R. J., *Mod. Plast.*, **41,** 149 (1964)
17. FISCH, W. and HOFMANN, W., *J. Polym. Sci.*, **12,** 497 (1954)
18. FISCH, W., HOFMANN, W. and KOSKIKALLIO, J., *J. appl. Chem. Lond.*, **6,** 429 (1956)
19. DEARBORN, E. C., FUOSS, R. M. and WHITE, A. F., *J. Polym. Sci.*, **16,** 201 (1955)
20. FISCHER, R. F., *J. Polym. Sci.*, **44,** 155 (1960)
21. TANAKI, Y. and KAKIUCHI, H., *J. appl. Polym. Sci.*, **7,** 1063 (1963)
22. DELMONTE, J., *J. appl. Polym. Sci.*, **2,** 108 (1957)
23. WARFIELD, R. W. and PETREE, M. C., *S.P.E. Trans.*, **1,** 3 (1961)
24. MILLER, B., *J. appl. Polym. Sci.*, **10,** 217 (1966)
25. DANNENBERG, H., *S.P.E. Jl.*, **15,** 875 (1959)
26. LEWIS, A. F. and GILHAM, J. K., *J. appl. Polym. Sci.*, **6,** 422 (1962)
27. JENKINS, R. and KARRE, L., *J. appl. Polym. Sci.*, **10,** 303 (1966)
28. LEWIS, A. F., *S.P.E. Trans.*, **3,** 201 (1963)
29. GOUGH, L. J. and SMITH, I. T., *J. appl. Polym. Sci.*, **3,** 362 (1960)
30. KAKURAI, T. and NOGUSHI, T., *Chemy. high Polym.*, **20,** 213, 21 (1963)
31. VAN HOORN, H. and BRUIN, P., *Paint, Varnish Prodn.*, **49,** 47 (1959)
32. EDWARDS, G. R., *Br. Plast.*, **33,** 203 (1960)
33. HARRIS, J. J. and TEMIN, S. C., *J. appl. Polym. Sci.*, **10,** 523 (1966)

CHAPTER 4

Curing Agents

4.1 INTRODUCTION

In the twenty years of epoxide resin technology, a vast number of compounds has been screened for their suitability as curing agents. Many compounds used in the early years of the technology have now been superseded by more sophisticated materials, though some still retain their popularity and even occasionally enhance it. This Chapter considers some of the curing agents that are either widely accepted and used in practical applications or have special features worthy of note.

The choice of curing agent to be used with an epoxide resin will depend upon:

(a) the handling characteristics required or tolerable in the un-cured system, such as viscosity at working temperature, pot life, exotherm, and toxicity,

(b) the cure and post-cure time and temperature requirements,

(c) the properties (physical, mechanical, electrical, and chemical) required of the cured system, and

(d) the cost of the curing agent.

The correct choice of curing agent can therefore be as important as the choice of the resin itself, both playing a part in determining the extent and nature of the intermolecular cross-linking.

Curing agents have been considered in the following broad categories:

(i) *Primary and secondary aliphatic polyamines and derivatives.* A group of unmodified room-temperature curing agents which can be regarded as having low viscosity and low cost, and being convenient (although not always pleasant) to use. Polyamides are also included; and they offer, in addition to room-temperature curing, the absence of unpleasant vapours and the formation of tough, flexible, cured products.

(ii) *Modified primary and secondary aliphatic amines*. Again room-temperature curing systems, developed to improve certain characteristics such as longer pot life, faster cures, and easier handling.

(iii) *Aromatic amines*. Essentially hot-curing systems, offering improved heat and chemical resistance and better strength properties.

(iv) *Acid anhydrides*. A family of long pot-life, low viscosity and low reactivity (in the absence of a catalyst), which provides cured systems equal in performance to the aromatic amines.

(v) *Catalytic curing agents*. Usually offer special properties, and frequently have long pot lives at room temperature.

The use of other resins as co-reactants with epoxides, such as phenol-formaldehyde and urea-formaldehyde resins, is considered in Chapter 8.

4.2 PRIMARY AND SECONDARY ALIPHATIC AMINES AND DERIVATIVES

Apart from phthalic anhydride, the amines were the first group of compounds to be widely used as cross-linking agents for epoxide resins. Naturally, amines in commercial production have been given priority in the screening process for potential curing agents and a number have been shown to give acceptable properties when used to cure resins (Fig. 4.1). However, wider usage of the resins has frequently been dependent upon achieving certain performance properties in the cured system. This has led to the deliberate synthesis of amine derivatives or the use of mixtures of amines, aimed at meeting specific user requirements.

In general, linear and branched primary and secondary aliphatic amines give with the glycidyl ether resins good fast room-temperature cures, resulting in heat deflection temperatures in the range 70–110 °C, depending upon the resin used, the concentration of the amine, and the cure cycle.[10] The cyclic aliphatic amines require a high-temperature cure, and give properties in the cured casting similar to those obtained from the aromatic amines, e.g. heat deflection temperatures of 140–150 °C. Table 4.1 gives the range of

Table 4.1 RANGE OF PROPERTIES OF CURED CASTINGS BASED ON TYPICAL ALIPHATIC
POLYAMINES, MODIFIED POLYAMINES AND POLYAMIDES
RESIN: DIGLYCIDYL ETHER OF DPP (WPE 180–195)

	Polyamine and modified polyamines	Polyamides
Pot life (500 g; 23 °C)	20–40 min	1–3 hr
Cure schedule	4–7 days at room temperature or gel at 23 °C + 1–2 hr at 100 °C	7 days at room temperature or gel at 23 °C + 1–2 hr at 100 °C
Heat deflection temperature,* °C	70–110	40–60
Ultimate tensile strength, lb/in²	7–10 000	4 500–6 500
Young's modulus, lb/in²	4–500 000	2–350 000
Ultimate compressive strength, lb/in²	12–15 000	7–9 000
Ultimate flexural strength, lb/in²	12–15 000	7–9 000
Izod impact strength, ft–lb/in notch	0·4–0·5	1·0–1·2
Dielectric constant (50 Hz; 23 °C)	4–5	3·2–3·6
Power factor (50 Hz; 23 °C)	0·02–0·03	0·010–0·035
Volume resistivity (ohm cm; 23 °C)	10^{15}	10^{15}

*For definition and discussion, see Section 5.3.3

properties obtained in cured castings using typical linear aliphatic
polyamines, modified polyamines, and polyamides. It does not
include the cycloaliphatic curing agents, which are dealt with in
Table 4.2 together with the aromatic amines. Aliphatic amines are
not widely used with the non-glycidyl ether resins, since the amine/
epoxide reaction is here slow at atmospheric temperatures and
usually requires heat and accelerators to obtain an acceptable rate
of cure.

The resin and the polyfunctional amine must interact in approxi-
mately stoichiometric amounts to obtain the best balance of
properties. In this stoichiometry it is assumed that the functionality
of the amino group towards the epoxide ring is one active amino
hydrogen for each epoxide ring. Consider curing agent DTA
(diethylenetriamine) as an example (Fig. 4.1). It has M 103 and five
active hydrogen atoms per molecule. Hence, the amine equivalent
weight per active hydrogen is $103/5 = 20·6$. If the resin has an
epoxide equivalent of 180, then 180 g will require 20·6 g of amine to
provide one hydrogen atom for each epoxide group. Expressed as
parts curing agent per 100 parts of resin (phr), the amount of amine
becomes $20·6 \times 100/180 = 10·2$.

The stoichiometric amounts for high-functionality amines are

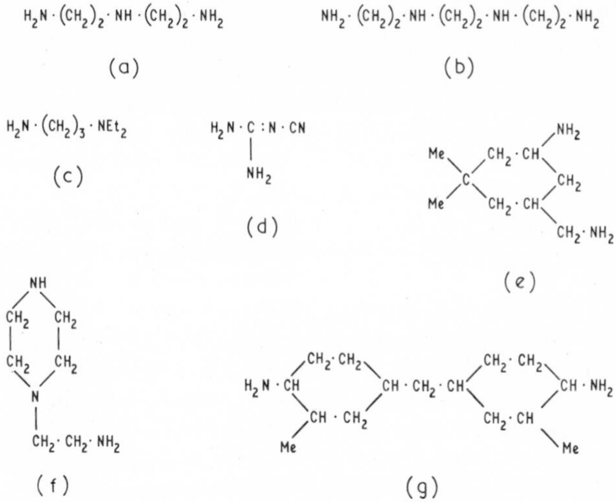

$$H_2N \cdot (CH_2)_2 \cdot NH \cdot (CH_2)_2 \cdot NH_2$$

(a)

$$NH_2 \cdot (CH_2)_2 \cdot NH \cdot (CH_2)_2 \cdot NH \cdot (CH_2)_2 \cdot NH_2$$

(b)

$$H_2N \cdot (CH_2)_3 \cdot NEt_2$$

(c)

$$H_2N \cdot C : N \cdot CN$$
$$| $$
$$NH_2$$

(d)

(e)

(f)

(g)

Fig. 4.1 Some typical polyamine curing agents: (a) diethylenetriamine, (b) triethylenetetramine, (c) diethylaminopropylamine, (d) dicyandiamide (DICY), (e) isophorone diamine, (f) N-aminoethylpiperazine, (g) bis (4-amino-3 methylcyclohexyl)methane

therefore small, and this can lead to practical handling difficulties. Automatic dispensing equipment, now widely used for epoxide resin systems, is less prone to inaccurate metering if the amount of curing agent used is about the same as the amount of resin. To achieve this, amines have been converted into derivatives of higher molecular weight by reaction with mono- or polyfunctional glycidyl compounds.

Most of the aliphatic amines are skin sensitisers and can give rise to dermatitis if handled carelessly. Modifications of the amines have been developed to overcome this difficulty or to achieve lower exotherms, longer pot lives, and varying speeds of cure.

4.2.1 DIETHYLENETRIAMINE (DTA) AND TRIETHYLENETETRAMINE (TET)

These two polyamines are pungent low-viscosity liquids, widely used with DPP–ECH resins for fast cures or where room temperature curing is required. The high rate of cure is accompanied by a very short pot life, together with rapid evolution of heat. These characteristics are limiting factors to the use of DTA and TET, and in particular restrict the size of castings that can be made.

Possessing M 103 and five active hydrogen atoms per molecule, DTA should be used at a concentration of about 10–11 phr with a liquid resin of WPE 175–210, for the full stoichiometric value to be obtained. Similarly TET, with M 150 and six hydrogen atoms per molecule, has a stoichiometric value of 14 phr. In practice the recommended amounts are 8–12 phr for DTA and 10–14 phr for TET. The use of reduced amounts is a compromise between an acceptable pot life and a reasonable level of properties in the cured casting.

Good all round properties are obtained from DTA or TET cures, but these values are not maintained above 50 °C. For maintenance of properties at higher temperatures, aromatic amines are preferred to aliphatic as curing agents. Typical cure schedules for DTA and TET systems are 4 days at room temperature (25 °C), or 1 hr at 100 °C. These curing agents are usually employed in laminating, adhesive, and certain surface-coating formulations.

4.2.2 DIETHYLAMINOPROPYLAMINE (DEAPA) AND N-AMINOETHYLPIPERAZINE (AEP)

Both of these curing agents are of limited applicability, and are usually chosen for very specific reasons. DEAPA, whilst giving properties very similar to those obtained with DTA, is less reactive and therefore has a lower rate of cure and a much longer pot life. At room temperature, with a liquid glycidyl ether resin, and masses of 500 g, DEAPA gives a pot life of 140 min, as compared with 20–30 min for DTA.

DEAPA, having M 130, achieves polymerisation of epoxides through its two active amino hydrogens and also the tertiary amino group in the molecule. It can therefore be used in concentrations considerably less than the stoichiometric amount based on the hydrogen atoms alone, the recommended amount of DEAPA being 7 phr.

AEP is a cycloaliphatic amine possessing primary, secondary, and tertiary amino groups, although only the first two are involved in curing. It is a clear, high-boiling liquid, with handling and curing characteristics similar to those of DTA and TET, e.g. 20–30 minutes pot life at room temperature; but it is not able to cure sections thicker than $\frac{1}{8}$ in at room temperature. The chief advantage of AEP is that it produces castings with much better impact properties than either DTA or TET.

Optimum cures of liquid diglycidyl ethers are obtained using 20–22 phr and a typical cure schedule is to gel at room temperature followed by 1 hr at 200 °C. This usually produces castings with HDTs of 108–110 °C and Izod impact strengths up to 1·1 ft–lb/in notch, values which are two to three times greater than are obtained with DTA and TET.

4.2.3 LAROMIN C 260*

This substance, which is the cycloaliphatic primary diamine bis(4-amino-3-methylcyclohexyl)methane, is a colourless liquid of low viscosity, easily miscible with epoxide resins. When used to cure a liquid diglycidyl ether at elevated temperatures, light-coloured castings are produced which possess good high-temperature physical and electrical properties.

Laromin C 260 is used at a concentration of 33 phr, the mixture having a pot life (500 g at 25 °C) of 3 hr. A typical cure would be to gel at 80 °C followed by 2–4 hr at 150 °C, the cured system having a HDT of 155–160 °C. Properties are maintained at elevated temperatures and the system is suitable for the production of small castings where class B insulation is needed, i.e. under continuous service at temperatures up to 130 °C. In general, castings based on Laromin C 260 have mechanical properties approaching those obtained from aromatic amines (Table 4.2) and fully equivalent electrical properties.

4.2.4 DICYANDIAMIDE (DICY)

This solid, one of the first curing agents to be used with epoxide resins, is widely employed in adhesives, 'prepreg' laminating, and powder-coating formulations. It can exist in tautomeric forms (Fig. 4.2), and Levine[1] has suggested that at elevated temperatures

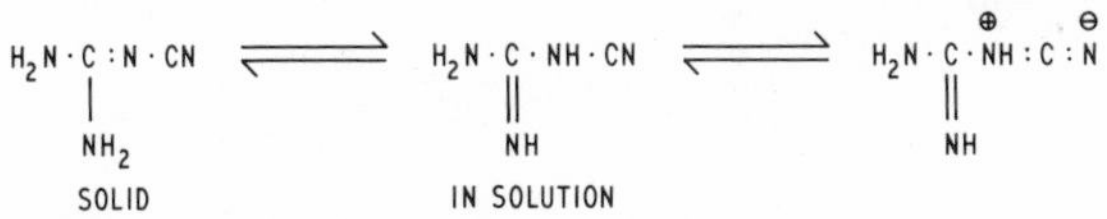

Fig. 4.2

*Badische Anilin u. Soda Fabrik A.G.

it behaves as a polyamine. However, since the imide group is present in the structure it is possible that other mechanisms can also occur.

DICY is used at a level of 4 phr and can be regarded as a latent curing agent with a pot life from one day to a year in certain circumstances. On heating to 145–160 °C a rapid polymerisation commences and acceptable properties are obtained after half an hour at that temperature.

4.2.5 POLYAMIDES

These curing agents are aminopolyamides, produced from the reaction of dimerised and trimerised vegetable oil fatty acids with polyamines.[2] The reaction is typified by the Diels-Alder reaction

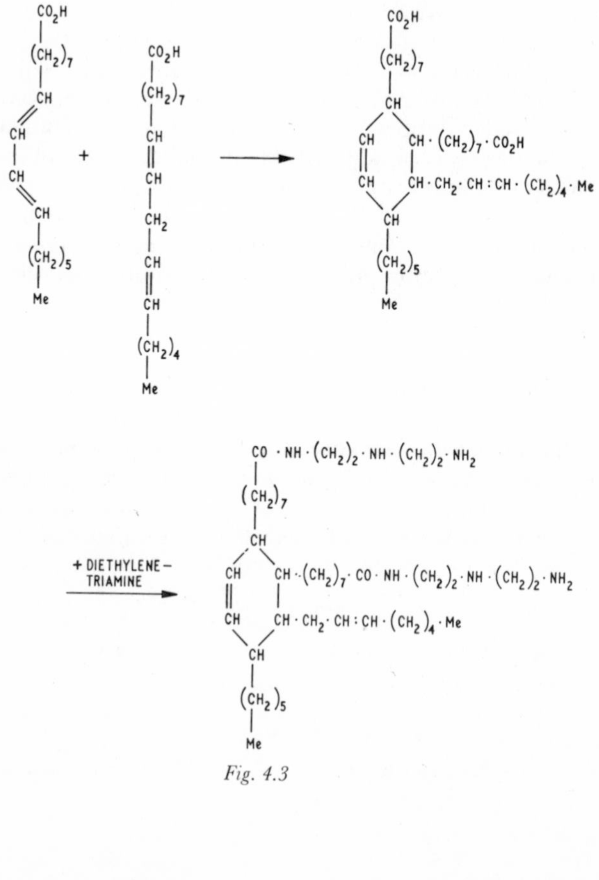

Fig. 4.3

between 9,12- and 9,11-linoleic acids, the resulting dimer forming with DTA two molecules of the aminopolyamide (Fig. 4.3). From the simplified structure of the polyamide, it can be seen that it contains not only amide but also primary and secondary amino groups, and it is the latter which are responsible for the curing reaction. In fact, the polyamides available commercially are complex mixtures of substances containing additionally free carboxyl groups and ring structures of the imidazoline type.[3-5]

A range of polyamides of this type is available from a number of suppliers. They were first introduced by General Mills Inc. under the trade-name Versamid, and the grades most used with epoxides are Versamid 115 and 125, the latter being the more reactive. Typical properties of these materials are:

	Versamid 115	*Versamid 125*
Amine value*	210–230	290–320
Amine equivalent weight	250	180
Viscosity; poises at 23°C	Solid	500
Specific gravity	0·99	0·97

*The *amine value* is an indication of the amount of hydrogen available for curing and is expressed in mg, KOH/g.

The polyamides are versatile curing agents for epoxides. They are not unpleasant to handle, are compatible with the resins, and their mixing ratios with resin are not critical. Fairly long pot lives are possible (about 4 hr), and mild cure schedules (such as 4 hr at 65°C for small masses) yield castings with good impact strength, excellent adhesion to many different substrates, and an HDT of about 60°C. Room-temperature curing is also possible, but this leads to a lower level of properties. Table 4.1 gives the broad range of properties obtained in polyamide-cured castings. The polyamides are regarded as flexibilisers as well as curing agents, this effect being achieved by the flexible nature of the molecule between the reactive points (Chapter 6).

4.3 MODIFIED PRIMARY AND SECONDARY AMINES

Many modifications have been made to the aliphatic amines in order to achieve:

(i) Improved characteristics such as longer pot life, faster or slower cures, resin compatibility, and lower readiness to carbonate in air.

(ii) Lower dermatitic potential.

(iii) Easier handling, lower viscosity, easier mixing by hand or machine when used in amounts similar to that of the resin.

The most commonly used modifications are those based on adducts of the polyamine with: (*a*) Ethylene oxide, (*b*) acrylonitrile, (*c*) diglycidyl ethers, and (*d*) other compounds containing groups reactive towards the amino group, such as ketones.

The properties obtained in castings cured with these modified polyamines are broadly similar to those of the unmodified amines, and a range of typical values is given in Table 4.1 (p. 62).

4.3.1 AMINE-ETHYLENE OXIDE ADDUCTS[6, 7]

Polyamines such as DTA react readily with ethylene oxide in the presence of water to yield the hydroxyalkyl derivatives, increasing concentration of the oxide leading to increasing substitution of the amino hydrogen atoms by the hydroxyethyl group (Fig. 4.4).

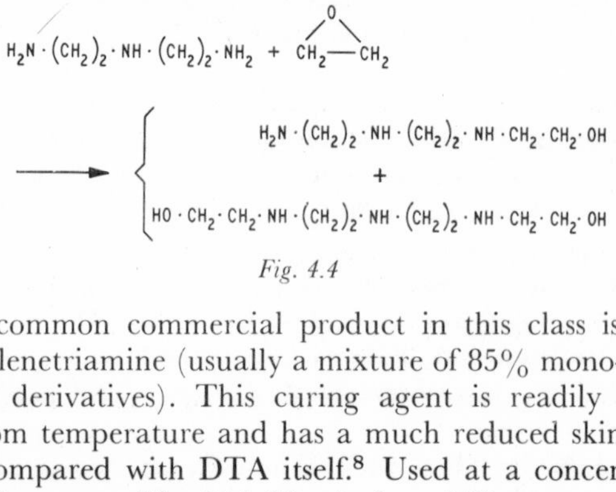

$$H_2N \cdot (CH_2)_2 \cdot NH \cdot (CH_2)_2 \cdot NH_2 \; + \; CH_2 \overset{O}{-\!\!\!-\!\!\!-} CH_2$$

$$\longrightarrow \left\{ \begin{array}{c} H_2N \cdot (CH_2)_2 \cdot NH \cdot (CH_2)_2 \cdot NH \cdot CH_2 \cdot CH_2 \cdot OH \\ + \\ HO \cdot CH_2 \cdot CH_2 \cdot NH \cdot (CH_2)_2 \cdot NH \cdot (CH_2)_2 \cdot NH \cdot CH_2 \cdot CH_2 \cdot OH \end{array} \right.$$

Fig. 4.4

The most common commercial product in this class is hydroxy-ethyldiethylenetriamine (usually a mixture of 85% mono- and 15% di-hydroxy derivatives). This curing agent is readily soluble in resin at room temperature and has a much reduced skin irritation potential compared with DTA itself.[8] Used at a concentration of 20 phr, it gives a pot life of 15–20 min for a 500 g mass, and a peak exotherm of about 200 °C.

Adducts of this type will provide fast room-temperature cures, but heat or a catalyst is often employed to accelerate the process. As with other curing agents the optimum conditions depend on the bulk of the mix and the amount of filler. Important cure schedules are: (*a*) 4 days at 23 °C (which gives a cured resin with fully developed mechanical properties, though a usually acceptable level of these properties is attained within a few hours of cure); (*b*) gel at 23 °C, followed by a post-cure of 2–4 hr at 65 °C or 1–2 hr at 100 °C, This cure schedule ensures full development of all properties.

4.3.2 AMINE-ACRYLONITRILE ADDUCTS[9, 10]

As with the ethylene oxide adducts, DTA is the preferred poly-amine for forming adducts with acrylonitrile (Fig. 4.5), the degree

$$H_2N \cdot (CH_2)_2 \cdot NH \cdot (CH_2)_2 \cdot NH_2 \ + \ CH_2 : CH \cdot CN$$

$$\longrightarrow \left\{ \begin{array}{l} H_2N \cdot (CH_2)_2 \cdot NH \cdot (CH_2)_2 \cdot NH \cdot CH_2 \cdot CH_2 \cdot CN \\ + \\ NC \cdot CH_2 \cdot CH_2 \cdot NH \cdot (CH_2)_2 \cdot NH \cdot (CH_2)_2 \cdot NH \cdot CH_2 \cdot CH_2 \cdot CN \end{array} \right.$$

Fig. 4.5

of substitution of hydrogen by the cyanoethyl group depending on the concentration of reactants and the reaction conditions.

The most important derivative is that formed from equimolecular amounts of DTA and acrylonitrile. When used at a concentration of 22 phr with a liquid diglycidyl ether resin, it gives a pot life of 40 min (500 g at 23 °C) and a peak exotherm of 170–180 °C as compared with DTA (240–250 °C).

Other advantages of this modification, apart from its reduced reactivity, are better wetting of glass fibre, and lower vapour pressures. Unfortunately there is no reduction in the skin-sensitising properties of DTA, as is achieved with the ethylene oxide adduct.

4.3.3 AMINE-RESIN ADDUCTS

Two types of adduct are important: those based on a liquid diglycidyl ether resin, and those based on a solid resin. The former type is a liquid, and the latter a solid which can be used as such but which is usually dissolved in a solvent mixture. Both types are prepared by allowing the resin to react with an excess of the amine, the resulting resinous 'amine adduct' containing residual amino hydrogens, all epoxide groups having been consumed (Fig. 4.6).

In a typical preparation of an adduct based on a liquid resin, 52% wt. DTA is warmed to about 75 °C in a stirred kettle provided with heating and cooling facilities. To the DTA is added 46% wt. of the resin, and the exothermic reaction resulting is controlled to

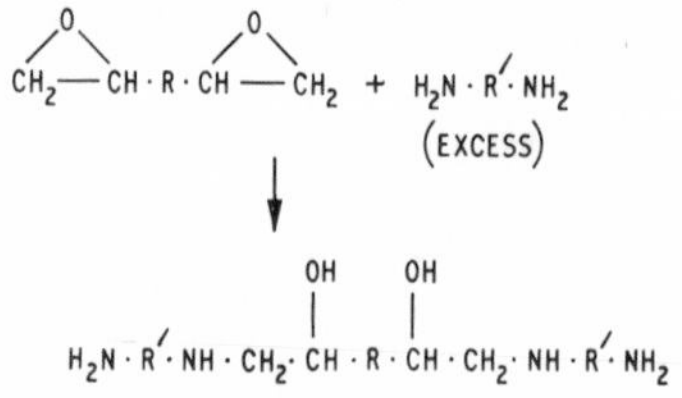

$$CH_2 \!-\! CH \cdot R \cdot CH \!-\! CH_2 \;\; + \;\; H_2N \cdot R' \cdot NH_2$$

(EXCESS)

$$H_2N \cdot R' \cdot NH \cdot CH_2 \cdot \overset{OH}{\underset{|}{CH}} \cdot R \cdot \overset{OH}{\underset{|}{CH}} \cdot CH_2 \cdot NH \cdot R' \cdot NH_2$$

Fig. 4.6

ensure a temperature of between 75 °C and 100 °C throughout the reaction. A cure accelerator is usually incorporated with the adduct, and in this case 2% wt. phenol is added at the end of the reaction, when the temperature has dropped to 55 °C. After filtration, the adduct is used unmodified.

Prepared as described above, the adduct will have a viscosity of 70–80 poises at room temperature and an amine equivalent of 45, and will be used at a concentration of 25 phr with a liquid resin. The adduct can be regarded as a modified fast-curing liquid amine, suitable for formulations where rapid room temperature cures are desirable. The advantages the adduct possesses over the conventional aliphatic polyamines can be summarised as follows:

1. Its lower vapour pressure minimises the amine odour problem.
2. It gives a more convenient resin-to-curing agent mixing ratio, which facilitates mixing and weighing.
3. Where extremely rapid cures are needed, the adduct gives better results than the unmodified polyamines.

A pot life of from 15–40 min is possible, depending upon mass and ambient temperature, and optimum properties are achieved at room temperature after 8–10 days. In general, castings prepared using the adduct are strong, tough, and shock-resistant, and are roughly comparable in properties to those obtained with DTA and TET cured systems.

Amine adducts derived from solid resins are prepared in a similar way to those based on liquid resins. The solid resin adducts find their main use in surface-coating systems and can be used either as an *in situ* adduct (the adduct not being isolated from the solvent solution in which it is prepared) or as an isolated adduct. In the latter case, the excess amine and solvents are removed by distillation at 150 °C at reduced pressure when the reaction has ended.

Two typical *in situ* adduct formulations are;

	Ethylenediamine adduct	*DTA adduct*
Epoxide resin (*M* 900)	35·0% wt.	32·6% wt.
Ethylenediamine (75% tech.)	6·2	—
DTA (pure)	—	7·4
n-Butanol	29·5	30·0
Toluene	29·3	30·0

The resin, as a 50% solution in the solvents, is added slowly to the amine in solution, with stirring, which is continued for up to 3 hr after the resin addition. The properties of surface-coatings using adduct cures are given in Chapter 8, together with the advantages of using adducts in place of free aliphatic polyamines.

4.3.4 KETIMINES[11, 12]

The reaction between ketones and primary aliphatic amines can lead to the formation of ketimines (Fig. 4.7).

$$2 \; \overset{R'}{\underset{R''}{\diagup}} C : O + H_2N \cdot R \cdot NH_2 \longrightarrow \overset{R'}{\underset{R''}{\diagup}} C : N \cdot R \cdot N : C \overset{R'}{\underset{R''}{\diagdown}}$$

Fig. 4.7

The ketimines can be regarded as 'blocked' amines, and since the $C = N$ bond of the condensate is readily hydrolysed to liberate the original amine and ketone, they are also latent curing agents. The reaction between two molecules of acetone and one of DTA leads to the formation of the imine:

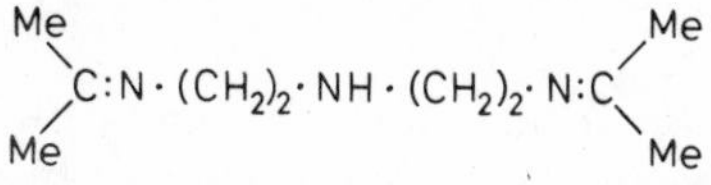

$$\overset{Me}{\underset{Me}{\diagup}} C : N \cdot (CH_2)_2 \cdot NH \cdot (CH_2)_2 \cdot N : C \overset{Me}{\underset{Me}{\diagdown}}$$

To remove the secondary amino hydrogen, the imine is allowed to react with phenyl glycidyl ether. The resulting compound can be used as a curing agent in high-solids surface-coating systems which have long pot lives and still cure at room temperature. The coatings can be applied in thicknesses up to 10×10^{-3} in, and when the system is exposed to moisture in the air, hydrolysis of the imino group occurs, liberating the amine which cures the coating. The ketone leaves the film by evaporation. Details of formulations using these blocked amines, and the properties of coatings based on them, are given in Chapter 8.

4.4 AROMATIC AMINES

Aromatic amines have been used since the earliest days of epoxide resin technology, chiefly because of the improved thermal and chemical resistance they impart to castings and glass cloth laminates, as compared with those obtained with aliphatic polyamines. In addition, aromatic amine-cured systems retain their properties better at elevated temperatures. They also have the advantage of forming dry brittle soluble solids when partially cured (B stage). These solids flow when heated, before finally forming a cured infusible network, and can therefore be used in the preparation of dry lay-up laminates.

At room temperature the aromatic amines react sluggishly with glycidyl ether resins, and to obtain acceptable properties elevated cures are necessary, sometimes in conjunction with an accelerator. The lower basicity of the aromatic amines, the steric factors operating in the reaction between the amino and epoxide groups, together with the relative lack of mobility of the growing polymer network all contribute to the difference in reactivity between aromatic and aliphatic amines.

The most important aromatic amine curing agents are:

m-Phenylenediamine (MPD)   4,4'-Diaminodiphenylmethane (DDM)   4,4'-Diaminodiphenyl sulphone (DDS)

All are solids at room temperature and can be difficult to mix with the resin. To overcome this difficulty, eutectic blends of MPD and DDM have been developed which are liquid at room temperature. Alternatively, the aromatic amine can be dissolved in a solvent such as dimethylformamide or γ-butyrolactone. These solutions can cure resin at room temperature, though naturally the properties of the castings are not comparable to those obtained from hot-cured systems.

4.4.1 *m*-PHENYLENEDIAMINE (MPD)[11]

This curing agent can be used with a liquid diglycidyl ether for the preparation of both castings and glass cloth laminates. In both uses, it provides a cured resin with excellent chemical resistance and retention of physical and electrical properties at elevated temperatures. Possessing four amino hydrogen atoms and M 108, MPD is used at a concentration of 14·5 phr with a liquid resin. To achieve thorough mixing it is usual for about 15% of the total resin to be heated to 65 °C, the molten MPD being added with stirring. After cooling to room temperature, the remainder of the resin is blended with the mix. Care should be taken to remove the irritating fumes of the amine evolved during this mixing operation, since aromatic amines, like the aliphatic, are skin irritants and sensitisers. MPD also has the unfortunate property of staining the skin upon contact.

When used at a concentration of 14·5 phr, 500 g batches have a pot life of 2·5 hr at 50 °C. Cure schedules depend upon the size of the casting being made, but it is usual for all cures to be in two stages. In this way the development of excessive exothermic heat can be avoided. A typical procedure is to gel the system at 85 °C for 2 hr and then post-cure for 4 hr at 150 °C.

Table 4.2 TYPICAL PROPERTIES OF CASTINGS CURED WITH AROMATIC AND CYCLOALIPHATIC AMINES
RESIN: DIGLYCIDYL ETHER OF DPP (WPE 185–190)

	Aromatic amine	*Cycloaliphatic amine*
Pot life (500 g; 23 °C)	8 hr	1–3 hr
Cure schedule	Gel at 80 °C + 4 hr at 150 °C	Gel at 80 °C + 4 hr at 150 °C
Heat deflection temperature, °C	145–150	145–150
Ultimate tensile strength, lb/in²	10–13 000	11–12 000
Ultimate compressive strength, lb/in²	17–19 000	—
Ultimate flexural strength, lb/in²	17–18 000	15–16 000
Izod impact strength, ft–lb/in notch	0·5–0·6	0·5–0·6
Dielectric constant (1 KHz; 25 °C)	4·0–4·5	3·8–4·3
Power factor (1 KHz; 25 °C)	0·013–0·017	0·015–0·019
Volume resistivity (ohm cm; 25 °C)	10^{16}	10^{16}

Typical physical properties of a liquid resin cured with an aromatic amine are given in Table 4.2. When MPD is used to cure high-functionality resins such as the epoxidised novolaks or mixtures of the novolak with a DPP–ECH resin, higher heat deflection temperatures (about 200 °C) are obtained, as expected.

4.4.2 4,4'-DIAMINODIPHENYLMETHANE (DDM)[12]

Resins polymerised with DDM have properties similar to those of polymers obtained with MPD, but have an advantage over the latter in that they do not stain the skin. DDM is used at a level of 27 phr, is mixed with resin in a similar way to MPD, and batches of 500 g have a pot life of 2·5 hr at 50 °C. A recommended cure schedule is to allow the mixture to gel at room temperature (8–12 hr) and then heat at 80–100 °C for two or more hours. A post-cure for one hour at 200 °C ensures optimum heat and chemical resistance.

DDM systems are ideally suited for electrical insulation, because of their excellent electrical and mechanical properties, these being maintained even under high humidity. Another popular use for the system is in wet and dry lay-up glass cloth laminating.

4.4.3 4,4'-DIAMINODIPHENYL SULPHONE (DDS)[13]

This powder has m.p. 176 °C, M 256, and four amino hydrogens. The stoichiometric amount to be used with a liquid diglycidyl ether resin is therefore 33·5 phr, but it is often found useful to increase this concentration to at least 36 phr to attain optimum properties. DDS reacts very slowly with resins even at high temperatures, probably owing to the effect of the sulphone group in decreasing the basicity of the amine. To shorten the cure time, 1 phr of the BF_3–MEA adduct can be added as an accelerator, and here 30 or even 20 phr of DDS is used. Because of its very high viscosity even at 25 °C, the mix is usually handled hot. At 130 °C, 500 g of the resin-DDS-accelerator system has a pot life of 0·5 hr, and when cured for 2 hr at 130 °C plus 1 hr at 200 °C gives a HDT of 180 °C.

Omission of the accelerator, plus a further hour at 200 °C, can give an HDT of 193 °C. In general, DDS gives the highest HDT values of all the amine curing agents and is preferred where high temperature applications are envisaged. Thus, when it is used to cure a 50:50 mixture of a liquid diglycidyl ether resin with a tetraglycidyl ether, an HDT of 226 °C is obtained.[11]

4.5 MODIFIED AROMATIC AMINES

To achieve mixing at room temperature, and hence a longer pot life, liquid eutectic mixtures of aromatic amines have been developed. The most popular contains DDM and MPD, and is readily miscible with liquid resins at room temperature. On standing, the eutectic deposits crystals which can be remelted by warming to 40 °C.

Certain solutions of aromatic amines will also polymerise resins at room temperature. The effect of the solvent is probably to allow sufficient mobility of the polymer chains for an adequate degree of cross-linking to occur before the cross-link density becomes so high that the molecules are now immobilised and further reaction stops. However, the incorporation of solvent in the cured resin has the effect of greatly lowering the HDT and the general level of properties associated with aromatic amines. These solutions give properties more akin to polyamide cures, but have the advantage of low viscosity and adjustable cure-rate, since they are usually used with a cure accelerator.

4.6 ACID ANHYDRIDES

These compounds form one of the most important groups of curing agents and were mentioned by Castan in his early work. Large numbers of anhydrides, including polymeric anhydrides, dianhydrides, and various compounds in combination with anhydrides, have been claimed in the patent literature as curing agents, but most of the important ones are based on a cycloaliphatic structure (Fig. 4.8).

When used with glycidyl ether resins, anhydrides provide cured systems that are light in colour and have good mechanical and electrical properties and better high temperature stability than the amine-cured systems. Typical physical properties of some anhydride cured systems are given in Table 4.3. The resin-anhydride mixture has a low viscosity, long pot life, and low volatility, and is non-dermatitic, although the breakdown products formed during cure may be irritating, or even toxic. During the hot cure there is only a little shrinkage of the system, and in general low exothermic heat evolution. The limitation to the use of anhydrides is the long and high temperature cure schedules needed to obtain optimum properties, although the use of catalysts (accelerators or promoters) assists in overcoming this drawback. Acid anhydrides are also the

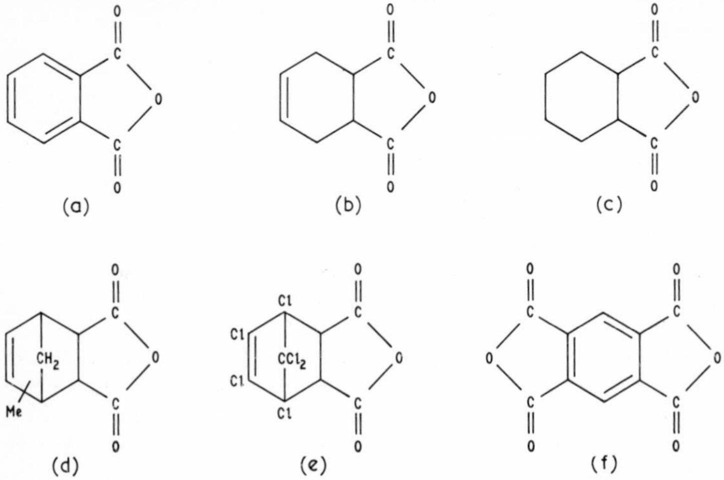

Fig. 4.8 (a) Phthalic anhydride (PA); (b) tetrahydrophthalic anhydride (THPA); (c) hexahydrophthalic anhydride (HPA); (d) nadic methyl anhydride (NMA)*; (e) chlorendic (HET) anhydride*; (f) pyromellitic dianhydride (PMDA) (* Trivial names adopted in the industry)

chief curing agents used for cross-linking cycloaliphatic and epoxidised olefin resins, which react more rapidly with anhydrides than with amines.

Table 4.3 PROPERTIES OF CASTINGS CURED WITH ACID ANHYDRIDES

	Phthalic anhydride/low M solid resin: WPE 330–380	HPA (80 phr) or NMA (90 phr); liquid resin WPE 180–195
Pot life (500 g; 23 °C)	60–90 min at 125 °C	4–5 days
Cure schedule	16–24 hr at 120–130 °C	2 hr at 80 °C + 5–15 hr at 125–150 °C
Heat deflection temperature, °C	100–110	125–135
Ultimate tensile strength, lb/in^2	12 000	12–13 000
Ultimate compressive strength, lb/in^2	16 000	17–19 000
Ultimate flexural strength, lb/in^2	18 000	18–19 000
Izod impact strength, ft–lb/in notch	0·7	0·4
Dielectric constant (50 Hz; 23 °C)	3·7	3·2–3·6
Power factor (50 Hz; 23 °C)	0·004	0·002–0·003
Volume resistivity (ohm cm; 23 °C)	10^{16}	10^{16}

It will be recalled from Chapter 3 that the anhydride group does not react very readily with the epoxide ring, whereas carboxyl and hydroxyl groups do so. Thus, in a simplified picture of the non-catalysed reaction, three steps can be distinguished:

1. Opening of the anhydride ring by a hydroxyl group, yielding a monocarboxylic ester.
2. Reaction of the carboxyl group of the monoester with the epoxide ring, forming a hydroxy diester.
3. Reaction of an epoxide ring with a hydroxyl, forming an ether link plus a new hydroxyl group.

Ester and ether links occur at equal frequency in cured resin, but a catalyst can change this balance, bases favouring esterification, acids favouring etherification. In uncatalysed systems, small amounts of free acid tend to promote etherification, and this consumes no anhydride. Hence, in uncatalysed systems, stoichiometric amounts of resin and anhydride are unnecessary. There is therefore no definite rule as to the proportions of anhydride to epoxide to be used, and the suggestions made by Arnold (Section 3.7) are the best guide. Naturally, to obtain an optimum value for a particular property there will be optimum values for cure schedule and amounts of anhydride and catalyst used; but these must be arrived at by experiment and will not necessarily optimise values for other properties.

The catalysts most frequently used, in amounts from 0·5 to 2·5 phr, are shown below.

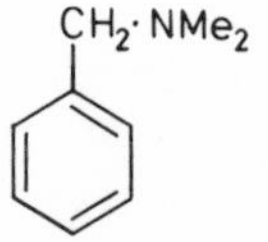

Benzyldimethylamine
(BDMA)

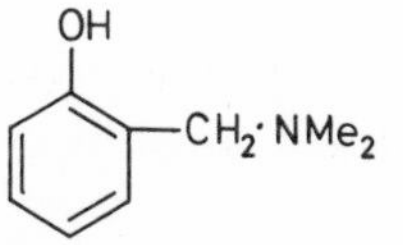

(Dimethylaminomethyl)phenol
(DMP-10)*

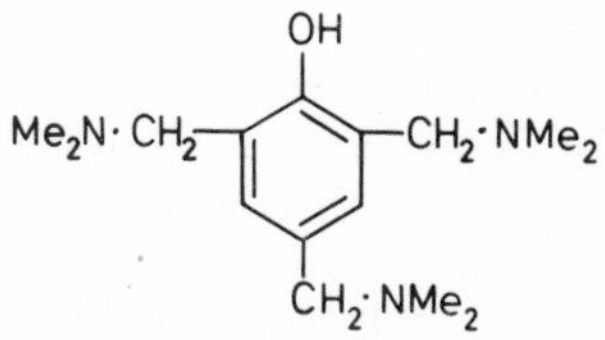

Tris(dimethylaminomethyl)phenol
(DMP-30)*

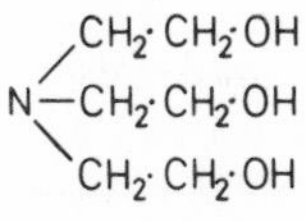

Triethanolamine

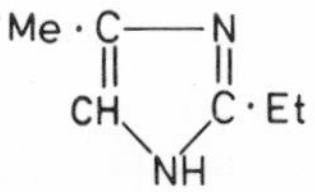

2-Ethyl-4-methylimidazole

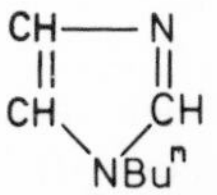

N-n-Butylimidazole

*Roehm and Haas Company.

These serve to speed gel times and reduce the length and temperature of cure. The use of low-M alcohols at 5 phr has also been
reported to reduce cure times,[14] both with the glycidyl ethers and
the cycloaliphatic resins.

4.6.1 PHTHALIC (PA), TETRAHYDROPHTHALIC (THPA), AND HEXAHYDROPHTHALIC (HPA) ANHYDRIDES

Phthalic anhydride, one of the first to be used with epoxide resins,[15]
is a solid (m.p. 128°C) which readily sublimes from hot mixtures
and is not easily soluble in resins. Its chief use is in conjunction with
a low-M solid glycidyl ether resin for the production of large
electrical castings, since it is relatively cheap and the system has
good electrical and crack-resistant properties. This anhydride is best
mixed by stirring it into the hot resin at 120°C, and when it has
dissolved, cooling the mix to 60°C to lengthen the pot life. Care
must be taken to ensure that the temperature does not drop below
60°C with consequent precipitation of the anhydride. PA is used
with the solid resin at 30 phr, usually without an accelerator, and
is cured at 120–130°C for 16–24 hr.

Tetrahydrophthalic anhydride has the advantage over PA of not
subliming, and is frequently lower in price. It is used to produce
light-coloured laminates and is mixed with resin at 80–100°C.
Below 70°C, THPA is precipitated and operations must therefore
be carried out above this temperature.

Hexahydrophthalic anhydride is the most versatile of the three
anhydrides, and is usually employed when light-coloured castings
or translucent laminates are required. It is a hygroscopic, low-
melting point solid (m.p. 35°C), which with liquid resins gives
mixes with long pot life and low viscosity at room temperature,
especially suitable for impregnating and laminating. Because of its
low reactivity, HPA is always used with an accelerator, usually
BDMA or DMP–30.

Casting systems of HPA with liquid resins show good thermal
shock resistance and have stable electrical properties up to 130°C,
hence their use for casting and potting in the electrical industry.
HPA mixes easily with resin at 50°C and is used at a concentration
between 55 and 80 phr depending on the nature of the resin.
Typical cure schedule for the liquid resin/HPA system plus 0·5–2·0
phr BDMA is 2 hr at 80°C plus 1 hr at 200°C.

4.6.2 NADIC METHYL ANHYDRIDE (NMA)

Nadic methyl anhydride (the accepted trivial name for a mixture of the isomers of methyl*endo*methylenetetrahydrophthalic anhydride) is the most versatile of all the anhydrides, since the balance of properties of the final casting can be varied over a very wide range by altering the resin-to-curing agent ratio, changing the type and concentration of the accelerator, and modifying the cure schedule. In general, the highest heat resistance and hardness are obtained by using stoichiometric amounts of NMA and long cure schedules at high temperatures. Decreasing the amount of anhydride and the cure temperature leads to an improvement in toughness and crack resistance, but to a reduction in heat resistance of the casting.

NMA is a liquid of viscosity 2 poises at 25°C, readily soluble in resins and therefore requiring no special mixing or handling techniques. The maximum amount of NMA usually employed is 90 phr, and the minimum about 60 phr, when used in conjunction with a liquid diglycidyl ether. At 90 phr and with no accelerator present, a 500 g batch at 23°C has a pot life of 2 months; incorporation of 0·5 phr of DMP–30 in the mix reduces this to 4–5 days.

As has been mentioned, wide variations in cure schedule are possible. To attain the highest heat resistance (e.g., HDT *c.* 210°C) a cure of 2 hr at 220–260°C is necessary. A more usual cure, using 90 phr NMA plus an imidazole accelerator, is 2 hr at 80–100°C plus 4 hr at 140–150°C. For reduced amounts of NMA, and to obtain castings with maximum toughness but with markedly reduced HDTs (e.g. 107°C) post-cures of 16–20 hr at 125°C are normal.

4.6.3. CHLORENDIC ANHYDRIDE (HET* ANHYDRIDE)

Chlorendic anhydride, like NMA, has a condensed ring structure, but has in addition six chlorine atoms, which confer flame-retardant properties on the cured resin. Moreover, HET-cured systems have excellent electrical and mechanical properties. This curing agent is therefore recommended for electrical castings and laminates which need to be flame-retardant or to be used in service at temperatures below 180°C.

HET anhydride is a white crystalline powder (m.p. 230°C), which is hygroscopic and rapidly hydrolysed to the acid. The free

*Hooker Chemical Corporation.

acid speeds cure, shortens pot life, lowers HDT, and is less soluble in resin than the anhydride itself. It is therefore essential to keep HET anhydride dry when in storage.

The anhydride is used at a concentration of 117 phr, usually without an accelerator, and is dissolved in resin heated to 100–120°C. The systems are usually cured for a minimum of 6 hr at 160°C. Increase in time after this does not change the HDT significantly, but a decrease in cure temperature below 160°C leads to a considerable drop in HDT.

4.6.4 PYROMELLITIC DIANHYDRIDE (PMDA)[16, 17]

This anhydride is a solid (m.p. 286°C) and contains two anhydride groups symmetrically attached to a benzene ring. This compact molecule, the simplest of the aromatic dianhydrides, can be used to achieve high cross-link densities, and hence high HDTs and good solvent-resistance.

The anhydride is insoluble in resin at room temperature, and is also very reactive towards the epoxide group. Thus, high temperature mixing techniques would merely lead to gelation of the mixture. To overcome this difficulty, four techniques have been developed:

(a) The reactivity of PMDA may be reduced by replacing a proportion of it with a monofunctional anhydride (usually maleic but sometimes phthalic). This blend can then be mixed with resin at 70°C. Heating is usually continued with stirring until the anhydrides are dissolved, the mix then being cooled to 90°C to extend the pot life. By this method up to 65% of PMDA may be used in the blend. Above this level the pot life becomes too short for practical purposes.

(b) PMDA may be dissolved in acetone at reflux temperature; the solution thus prepared is stable for 7 days and can be used for pre-preg laminating.

(c) On interaction of PMDA with a glycol, a resin soluble adduct is obtained,[18] which can be used in adhesives and coatings.

(d) The PMDA can be suspended in liquid resin at room temperature, the subsequent hot cure promoting solution followed by reaction. When this method is used, reduced amounts of PMDA are preferred, *e.g.* anhydride-to-epoxide ratios of 0·4:1–0·5:1.

4.7 ANHYDRIDE BLENDS

Eutectic mixtures (of two anhydrides) which are liquids at room temperature are sometimes used to overcome difficult mixing procedures, and when HET is one of the components flame-retardancy is imparted to the casting. Some typical blends are:[11]

Blend	Ratio	phr	Accelerator (phr)	Approx. HDT (°C)
HET/HPA	2:3	96	1·0	130
HET/NMA	2:3	102	1·0	120
NMA/HPA	2:3	85	1·0	110

4.8 CATALYTIC CURING AGENTS

All of the curing agents so far described have achieved cure primarily by a polyaddition reaction. The curing agents are (or, as with anhydrides, are converted into) compounds that have labile hydrogen atoms which react on a 1:1 basis with the epoxide group. The resulting cross-linked network has the curing agent built into it as the means of holding the resin molecules together.

Catalytic curing agents achieve cross-linking by initially opening the epoxide ring and causing homopolymerisation of the resin. The resin molecules react directly with each other and the cured polymer has essentially a polyether structure. Although both Lewis acids and bases will bring about this type of cross-linking, the mechanisms of cure differ (see Chapter 3). The amounts of catalysts used with epoxide resins are usually determined empirically and are chosen to give the optimum balance of properties under the recommended working conditions. Catalytic curing agents can be used in any of three ways—as a sole curing agent, as a co-curing agent in conjunction with a polyamine or polyamide, or as an accelerator for anhydride systems.

4.8.1 LEWIS BASE CATALYSTS

Into this class fall the tertiary amines, the most widely used of the catalysts. Some metal alkoxides, such as aluminium isopropoxide and alkali metal hydroxides, also polymerise epoxide resins via an anionic mechanism. Monofunctional secondary amines also function as Lewis base catalysts once their active hydrogen atoms have been consumed by reaction with an epoxide group. Typical curing agents of this type are the imidazoles. Two very widely used tertiary amines are *o*-(dimethylaminomethyl)phenol (DMP–10)* and tris-(dimethylaminomethyl)phenol (DMP–30)* and its salt with 2-ethylhexoic acid (which is known as K61B).

DMP–10 and DMP–30 are used at concentrations of 4–10 phr with a liquid diglycidyl ether and achieve fairly fast cures overnight, even at 25°C, the hydroxyl group present in the molecule enhancing the catalytic activity of the tertiary amino groups. Curing agent K61B, used at 10·5 phr, gives a fairly long pot life, i.e., 7 hr at 20°C for 500 g, coupled with a moderate cure schedule—2 hr at 80°C. The acid moiety blocks the tertiary amine centres and deactivates them. The salt dissociates on heating, freeing the amine groups which are then able to react with an epoxide group. The phenolic

Table 4.4 PROPERTIES OF CASTINGS CURED WITH CATALYTIC CURING AGENTS (RESIN: DIGLYCIDYL ETHER OF DPP; WPE 180–195)

	K61B (10–12 phr) or Imidazole (4 phr)	*BF$_3$–400 (4 phr)*
Pot life (500 g; 23°C)	6–10 hr (15–40 min at 60°C)	about 4 months (3 days at 65°C, 3 hr at 100°C)
Cure schedule	4–8 hr at 60°C	4 hr at 115°C + 4 hr at 200°C
Heat deflection temperature, °C	70–130	176
Ultimate tensile strength, lb/in^2	8–11 000	—
Ultimate compressive strength, lb/in^2	13–14 000	—
Ultimate flexural strength, lb/in^2	17–18 500	—
Izod impact strength, ft–lb/in notch	0·3–0·4	—
Dielectric constant		
50 Hz; 23°C	3·6–3·9	—
1 MHz; 23°C	3·2–3·5	3·9
Power factor:		
50 Hz; 23°C	0·004	—
1 MHz; 23°C	0·022	—
Volume resistivity (ohm cm; 23°C)	10^{15}	10^{15}

salt is not a true latent curing agent since the free hydroxyl group can polymerise epoxide resins. Properties of typical K61B-cured systems are given in Table 4.4. Two other tertiary amines, used chiefly as catalysts for the anhydride-epoxide reaction, are benzyl-dimethylamine and α-methylbenzyldimethylamine. Substituted

*Roehm and Haas Company.

imidazoles are also useful curing agents, the most important being 2-ethyl-4-methylimidazole, a liquid of viscosity 40–80 poises at 25 °C. This substance can be used as a curing agent by itself, at a concentration of 10 phr, or as a catalyst for anhydride systems at 1 phr.

When used as a curing agent for a liquid resin, long pot lives (8–10 hr for 500 g at 25 °C) are obtained, coupled with a moderate cure schedule, 6–8 hr at 60 °C. The cured resin has an HDT of between 85–130 °C, which is increased by a post-cure to about 160 °C. Compared with other curing agents which polymerise epoxides under similar conditions, the imidazole offers improved HDTs and retention of mechanical properties at elevated temperatures.

Farkas and Strohm[19] have investigated the mechanism of cure and found that the imidazole becomes permanently attached to the polymer chain. They suggested the sequence shown in Fig. 4.9.

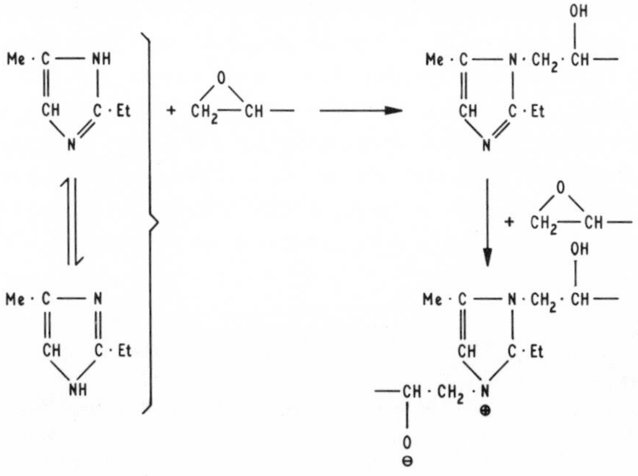

Fig. 4.9

The imidazole is thus an effective cross-linking agent, operating through both the secondary amino hydrogen and the tertiary amine, in a catalytic mechanism.

4.8.2 LEWIS ACID CATALYSTS

Lewis acids are electron pair acceptors, and include aluminium chloride, zinc chloride, stannic chloride, titanium tetrachloride, and boron trifluoride, all of which have some applicability to epoxide resin technology. They function as curing agents by coordinating with the epoxide oxygen (Chapter 3), facilitating the transfer of a proton, and they are effective catalysts for the polymerisation of linear and cycloaliphatic epoxides as well as for the glycidyl ethers. The most important of these compounds is the corrosive gas, boron trifluoride. This substance reacts extremely rapidly with the epoxide group at room temperature, polymerisation commencing in as short a period as 30 sec. In order to reduce this activity a variety of blocking techniques is possible, but the only one of any significance is the formation of a complex between the boron trifluoride and an amine such as monoethylamine. This complex (m.p. 90 °C), known commercially as BF_3:400, is a latent catalyst, which is hygroscopic and hydrolyses in moist air.

When used at 3–4 phr concentration, BF_3:400 mixed with a liquid glycidyl ether has a pot life of about 4 months at room temperature. On heating to 100–120 °C, the complex dissociates, freeing the BF_3, and cure commences. The rate of cure is very sensitive to temperature; below 100 °C the rate is negligible but at 120 °C a very rapid reaction occurs, accompanied by a considerable evolution of heat and a consequent large rise in the temperature of the casting. Care must therefore be taken when using BF_3:400 to cure castings over 100 g in weight, and large castings should not be attempted with this curing agent. The principal use for the complex is therefore as a catalyst with other curing agents (e.g., acid anhydrides) or in the preparation of laminates.

REFERENCES

1. LEVINE, H. H., *Am. chem. Soc. Symp.*, 148th meeting (1964)
2. RENFREW, M. M. and WITTCOFF, H., U. S. Pat. 2,705,223
3. PEERMAN, D. E., TOLBERG, W. and WITTCOFF, H., *J. Am. chem. Soc.*, **76,** 6085 (1954)
4. PEERMAN, D. E., TOLBERG, W. and WITTCOFF, H., *Ind. Engng. Chem.*, **49,** 1091 (1957)
5. TAWN, A. R. H., Paper to VIth FATIPEC congress, Wiesbaden (1962)
6. INGBERMAN, A. K., PITT, C. F. and PAUL, M. N., *Ind. Engng. Chem.*, **49,** 1105 (1957)
7. INGBERMAN, A. K. and WALTON, R. K., *J. Polym. Sci.*, **28,** 468 (1958)
8. PITT, C. F. and PAUL, M. N., *Mod. Plast.*, **34,** 125 (1957)
9. FARNHAM, A. G., U.S. Pat. 2,753,323
10. ALLEN, F. J. and HUNTER, W. M., *J. appl. Chem. Lond.*, **7,** 86 (1957)
11. Shell Chemical Co., Technical Literature
12. Shell Internationale Research Maatschappij, Brit. Pat. 905,725
13. MCGROARTY, J. A., *Ind. Engng. Chem.*, **52,** 1, 17 (1960)
14. N.V. de Bataafse Petroleum Maatschappij, Brit. Pat. 792,237
15. CASTAN, P., Brit. Pat. 518,057
16. E. I. Du Pont de Nemours & Co., Technical Literature
17. FIELD, R. B. and ROBINSON, C. F., *Ind. Engng. Chem.*, **49,** 369 (1957)
18. E. I. Du Pont de Nemours & Co., Brit. Pat. 886,601
19. FARKAS, A. and STROHM, P. F., *J. appl. Polym. Sci.*, **12,** 159 (1968)

CHAPTER 5

Characteristics of the Cured Resin

5.1 INTRODUCTION

In a previous Chapter the nature of the epoxide resins was considered, and it was established that each commercial grade of resin based on DPP and ECH consisted of a mixture of molecules of different chain lengths and therefore of different molecular weights, there being on average more than one epoxide group per molecule. The conversion of these molecules to a cross-linked network (i.e., the curing of the resins) was then reviewed, together with a number of the more important curing agents used to achieve this cross-linking.

This Chapter is concerned with the characteristics of the three-dimensional network, that is, the cured resin. It is clear that these characteristics, physical, electrical, chemical, all stem from the basic molecular structure of the polymerised resin. Important factors at the molecular level in determining these properties are:

   (i) the extent of cross-linking, i.e., the degree of cure,
   (ii) the cross-link density,
   (iii) the nature of the resin molecule between cross-links, and
   (iv) the nature of the curing agent molecule.

To obtain the most favourable properties in any cured resin system it is important to achieve maximum cross-linking. Upper limits to this may, however, be set by the trapping of reactive molecules in the polymer network at the gel stage, which are thus prevented from reacting. This condition can often be improved by a post-cure at a temperature above the original cure temperature and above the glass transition temperature for the system.[2] The increased molecular mobility brought about by heating gives the immobilised molecules a further opportunity to undergo collision and bond formation. A small proportion (2–3 phr) of a non-reactive diluent may also serve to improve molecular mobility and lead to a greater degree of cross-linking, e.g., the use of lactones in conjunction with aromatic diamines; but properties are down-graded because of the external plasticising effect of the diluent.

Molecules present in the resin which do not have one or more terminal epoxide group and hence do not enter into the curing reaction also lead to lower degrees of cure. Modern resin manufacture has now minimised the amount of these non-epoxide impurities in commercial grades of resin (Chapter 2).

A measure of the degree of cure is often taken to be the extent of conversion, *i.e.* the extent to which epoxide groups are consumed. Care must be taken when making this assumption, since the loss of an epoxide group does not necessarily mean that a bond has been formed contributing to further cross-linking. Dannenberg and Harp[3] examined three systems, each based on a commercial grade of the diglycidyl ether of DPP, cured with (*a*) piperidine, (*b*) DMP–30, and (*c*) MPD. Conversion of epoxide groups was measured by infra-red spectroscopy and chemical analysis, and compared with the extent of cross-linking as determined by solvent swelling and softening temperature. They found that the catalytic curing agents, at all curing temperatures considered but especially at higher temperatures, converted a higher percentage of epoxide groups than did MPD. However, despite its lower conversion of epoxide, the aromatic amine gave a cured product with the best solvent resistance and highest softening point.

The extent to which conversion of epoxide groups occurs with four different curing agents at 25 °C is illustrated in Fig. 5.1. After 3 weeks at room temperature, about 60% conversion had been attained except in the case of the polyamide cure when only 35% of the epoxide groups had reacted.

The cross-link density in different systems depends upon the functionality of the reacting species and the distance between their reactive groups. The nature of the resin or curing agent molecule between reactive groups, whether it is rigid or flexible, also has a direct influence on physical characteristics.

Resin systems with high cross-link densities normally exhibit higher HDTs and glass transition temperatures (Section 5.2) than those with more widely spaced cross-links, and it was the need for increasingly temperature-resistant systems that led to the development of resins and curing agents with increased functionality,

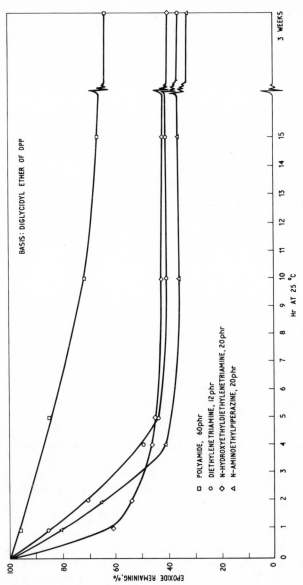

Fig. 5.1 Effect of structure on degree of cure[1]

such as epoxidised novolaks, the tetraglycidyl ether of tetraphenylol-ethane, and PMDA. Cross-link density is increased not only by increasing the functionality of the reacting species but also by decreasing the distance between the functional groups. This leads to improvements in short-term thermal stability, chemical resistance, and tensile and compressive strengths.

Decreases in cross-link density can be achieved by using a mono-functional 'chain-stopper' such as one of the mono-epoxide reactive diluents, or resins and curing agents with widely spaced functional groups. This has the effect of decreasing cure shrinkage and in some cases increasing toughness and elongation. A comprehensive study of the relationship between resin structure and physical properties by Burhans et al.[4] is discussed in Chapter 7.

5.2 NATURE OF THE GLASS TRANSITION TEMPERATURE (T_g)

As the temperature of a polymer is raised through its glass transition point the nature of the polymer changes from hard, glassy, and brittle to a softer and more flexible rubbery state. This is frequently accompanied by quite dramatic changes in properties such as refractive index, thermal conductivity, dielectric loss, mechanical stiffness and moduli, heat capacity, and the volume expansion coefficient. Traditionally, T_g has been obtained by measuring the specific volume of the glassy material as a function of tempera-ture. At the glass transition the specific volume-temperature curve has an abrupt change in slope.

Below T_g, molecular motion is frozen and there is not sufficient thermal energy to allow segments of the molecular chain to move as a whole. At T_g the increase in thermal energy has become sufficient to allow the movement of relatively large molecular segments which can jump from one position to another by rotation about carbon-to-carbon bonds.[5]

Glass transitions are exhibited by all but a few polymers. One exception is provided by highly cross-linked polymer systems (such as epoxides and some phenolics) where the polymers pyrolyse before they can change from the glass to the rubbery state, the bond dissociation energies being lower than the rotational barriers.[5] The glass transition temperature of a cured epoxide resin will therefore reflect the extent and nature of its cross-linking and can be used as a measure of its thermal stability. Because of its fundamental nature and the ease with which it is determined by differential thermal analysis or dynamic mechanical testing, T_g is replacing HDT as a method of evaluating and comparing different epoxide systems.

5.3 MEASUREMENT OF THE EXTENT OF CROSS-LINKING

Many different methods have been used to estimate the degree of cure of epoxide resins, including chemical analysis,[3] infra-red and near infra-red spectroscopy,[3,6,7] and the determination of HDT, glass transition temperature, hardness, electrical volume resistivity,[8] degree of swelling, and dynamic mechanical properties. The first two methods measure the extent to which the epoxide groups are consumed and are widely used as a measure of degree of cure. The other methods are based on the measurement of properties that are directly or indirectly related to the extent and nature of the cross-links.

5.3.1 CHEMICAL ANALYSIS

The swollen particle method described by Dannenberg and Harp[3] is the one usually adopted to determine the concentration of epoxide groups in a cured polymer. It consists of reducing the cured resin to a very fine particle size and suspending that powder in a solvent which acts as a swelling agent. The titration (see Section 2.4.3) is then carried out with these swollen particles, reagents diffusing into and out of the swollen particle after the reaction has been carried out. The values for epoxide content obtained by this method have been shown to be comparable to those obtained from infra-red spectroscopy.

5.3.2 INFRA-RED SPECTROSCOPY

The use of infra-red and near infra-red spectroscopy to measure epoxide concentration has already been described (Chapter 2). This technique is also applicable to the determination of the epoxide content of a resin system during and after cure.[3,6,7] Absorption spectra taken at various stages during cure can show not only the loss of epoxide over time but also the increase in hydroxyl concentration. Primary amine curing agents lead to a steady increase in hydroxyl concentration throughout the polymerisation, whereas tertiary amines cause an initial loss of hydroxyl followed by a notable increase towards the end of the reaction.

Infra-red spectroscopy has also been used by Harrod[9] to investigate the extent of hydrogen bonding in cured epoxide networks, using model epoxide compounds and amine-cured epoxide resins. He concluded that in the temperature range 30–200 °C the hydroxyl group is extensively involved in hydrogen bond formation. At the lower temperatures these bonds were chiefly thought to be of the long range type especially —O—H .... N<. Even at 200 °C short-range bonds occurred, relatively few free hydroxyl groups being present.

5.3.3 HEAT DEFLECTION TEMPERATURE (HDT)

The measurement of the *heat deflection temperature* of a cured resin is one of the most widely used methods of determining the degree of cure. Most HDTs are determined by the method outlined in ASTM D648–56. This states that the method determines the temperature at which an arbitrary deformation occurs when specimens are subjected to an arbitrary set of testing conditions. Data obtained in this way may only be used to predict behaviour of the test material at elevated temperatures, when time, temperature, method of loading, and fibre stress are similar to those in the test. In practice a bar of the cured resin is cast to specified dimensions, usually 5 in long by $\frac{1}{2}$ in wide and $\frac{1}{2}$ in thick, and supported at either end and loaded in the middle to produce a uniform stress of 264 psi. The sample is immersed in an oil bath with a deflection gauge mounted on the specimen. The temperature of the bath is raised at a rate of 2 degC per minute and the HDT is taken as the bath temperature at the point when the deflection of the bar has reached 0·010 in.

Clearly, the greater the extent of cross-linking in any one system, the higher the HDT. Comparison of HDTs and hence an assessment of degree of cure between various systems can, however, be misleading, since factors other than the extent of cross-linking may affect the HDT. Hoerner *et al.*[10] determined the HDT of a wide range of resin and curing agent combinations and pointed out that HDT cannot be readily correlated with other properties at elevated temperatures. Rather it can be used as a rough guide to the range of temperatures over which the various physical and electrical properties may be expected to remain constant or vary very little.

The HDTs of cured epoxide resins are generally in the range 50–200 °C, although certain highly functional systems such as those based on epoxidised novolaks can be made to have HDTs up to about 280 °C.

5.3.4 MEASUREMENT OF GLASS TRANSITION TEMPERATURE (T_g)

The traditional dilatometric method of measuring the glass transition point of a polymer has been used for epoxide systems. Kwei and Kummins[11] examined a liquid resin system cured with hexamethylenediamine plus 0–12% titanium dioxide filler or copper phthalocyanine pigment, and observed a T_g of 54 °C. The presence of filler or pigment made no difference to this value. Klopfenstein and Lee[12] give a typical thermal expansion curve for a cured liquid system which shows the abrupt break in linearity of the expansion curve at the glass transition point. Warfield[13] determined T_g by analysis of the temperature dependence of the electrical volume resistivity of three cured epoxide polymers and compared the values obtained to those derived from specific heat measurements. For the system employing K61B as the curing agent, resistivity measurements gave a value for T_g of 76 °C and specific heat measurements yielded 63–80 °C. Comparable values for the DEAPA-cured system were 110 °C and 60–115 °C.

Perhaps the most rapid and convenient method for T_g determination is differential thermal analysis (DTA). This technique, which has been fully described in the literature,[14–16] consists of heating the material under test and an inert reference material side by side under identical conditions and continuously recording the temperature difference (ΔT) between the two materials. Any exothermic or endothermic process occurring in the test sample will be recorded as a positive or negative peak in the curve (thermogram) of difference in temperature against temperature. DTA has been used to study the thermal degradation of epoxide resins (Section 5.5), the heats of polymerisation of epoxide resins (Chapter 3), and as a general guide to thermal stability.

Glass transition points for a series of epoxide resins cured with various amounts of polyazelaic polyanhydride (PAPA) were determined by Fava[17] using differential scanning calorimetry, which is a modified form of differential thermal analysis. Differential scanning calorimetry, instead of simply measuring the differential temperature caused by heat changes in the sample, measures the difference in rates of heat absorption by a sample with respect to an inert reference at each temperature, as the temperature is increased at a constant rate. In the absence of a chemical reaction a second order transition (glass transition) is shown as a discontinuity in the thermogram. The HDTs of the series of PAPA-cured resins, which possessed different softening points, were compared with their T_gs, and a good correlation was found between these two properties. Fava pointed out that T_g increases during cure and is a sensitive index of effective cure. Once T_g reaches the cure temperature, further reaction in the glassy state is extremely limited and virtually absent; hence the importance of curing a resin above its glass transition for optimum properties.

Glass transition temperatures can also be determined during the measurement of dynamic mechanical properties (see following section).

5.3.5 DYNAMIC MECHANICAL PROPERTIES

The measurement of dynamic mechanical properties is an attractive method for studying the cross-linking and structure of a cured resin beyond the area of Newtonian behaviour (Drumm et al.[18]). The technique is particularly valuable since it is more sensitive to changes in molecular structure than most mechanical tests. It also allows measurement over a wide range of temperatures with a single sample and provides more information in a single measurement than most other physical tests.

Dynamic mechanical tests measure the response or deformation of a material to periodic or varying forces. The applied force usually varies sinusoidally with time, as does the resulting deformation. Depending upon the method of testing, an elastic modulus (either Young's, shear, or bulk) can be obtained together with a mechanical damping. The mechanical damping (viscous or loss) factor is a measure of the energy dissipated each dynamic cycle and lost as heat. High polymeric materials are viscoelastic and have some of the properties of both elastic springs and viscous liquids. When they are deformed part of the energy provided is stored as potential energy and part is dissipated as heat, which manifests itself as mechanical damping. A valuable account of the theory and practice of dynamic mechanical testing is given by Nielsen.[19]

A cross-linked epoxide resin below its glass transition point is

Table 5.1 THERMAL PROPERTIES OF RESIN CURES* (AVERAGE VALUES)[20]

Property	Method	Cure 1	Cure 2	Cure 3	Cure 4	Cure 5
Heat deflection temperature, °C; 264 lb/in²	ASTM D648–45T	120	91	131	119	47
Modified Vicat penetration temperature, °C		125	93	136	123	51
Coefficient of linear expansion (in/in degC)	ASTM D696–44					
(a) Below T_g		6.38×10^{-5}	6.00×10^{-5}	5.83×10^{-5}	7.14×10^{-5}	7.79×10^{-5}
(b) Above T_g		16.4×10^{-5}	17.9×10^{-5}	20.8×10^{-5}	18.5×10^{-5}	20.0×10^{-5}
Glass transition, °C	ASTM D696–44	141	122	190	149	47

*Resin used: Diglycidyl ether of DPP
Cure 1: BF_3 : amine complex
Cure 2: DTA
Cure 3: MPD
Cure 4: Methylenebis(o-chloroaniline)
Cure 5: Aliphatic diamine

an amorphous rigid solid, having high modulus and low damping, both with low temperature-dependence. As the temperature is raised a slow decrease in modulus and increase in damping occurs. When the area of glass transition is reached the modulus decreases dramatically and the damping increases to a maximum (T_g). Above T_g the rubbery state is achieved, with low modulus and damping.

One of the first studies of dynamic mechanical properties of thermosetting resins was by Drumm *et al.*[18] who showed that these measurements could be used to study the degree of cure of phenolic resins. The cross-linking of glycidyl ether resins with various curing agents such as boron trifluoride complexes, DTA, and MPD was investigated by Kaelble,[20] using dynamic testing. Good correlation was established between the HDT, the Vicat penetration temperature, and the glass transition temperature obtained from dilatometry and dynamic mechanical measurements (Table 5.1).

Kline[21] studied the diglycidyl ether of DPP, cured with MPD, and also incorporated various amounts of aluminium powder filler into the system. Dynamic mechanical properties within the range 80–600 °C were measured, and the effects of cure time, filler content, and low dosage irradiation were investigated.

A torsion pendulum operating at a frequency of one cycle per second was used by May and Weir[22] to investigate the same resin system. They observed that the loss factor curve showed two transitions, a minor one at -60 °C, and the major one at 165–170 °C, some 15–20 degC above the HDT of the system as commonly observed. The structural origin of these peaks was then explored,

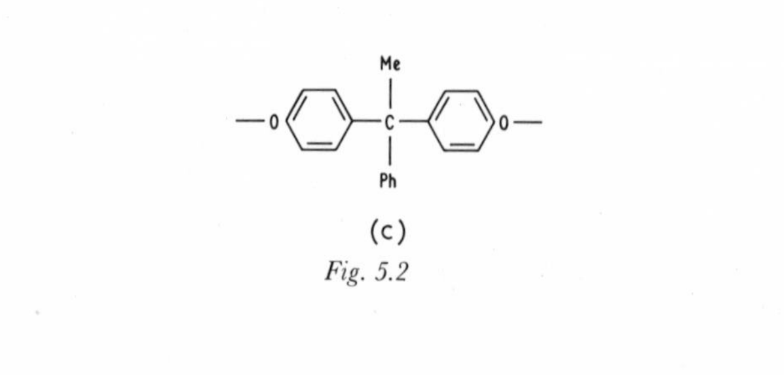

(a)   (b)

(c)

Fig. 5.2

first by measuring the dynamic mechanical properties of three resins each with different structures (Fig. 5.2) bridging the two aromatic rings in the resin chain.

The curves obtained indicated that the substituents on the central carbon bridge did not affect the low temperature peak but clearly affected the glass transition point, the value of T_g increasing in the order (a), (b), and (c) (Fig. 5.2). The low temperature peak was thought to be derived from the movement of small segments of the meshed molecular chains as in second order transitions in thermoplastics. To study this effect further, crystalline diglycidyl ether of DPP was cross-linked with three curing agents—a boron trifluoride-ethylamine complex, HPA, and PA. No low temperature peak was observed with the resins cured by the BF_3 complex and HPA, in contrast to the original MPD cured systems examined. It was therefore suggested that the low temperature peak was due to the existence of a sequence of three carbon atoms whose movement was not restricted except perhaps by hydrogen bond formation via the hydroxyl group, so that rotation of the segment was possible (Fig. 5.3). Structures (b) and (c) however, do not have a sequence of three or four unrestricted carbon atoms, free to rotate. The low temperature peak indicative of such a secondary transition is therefore absent. Anomalously, PA gave a low peak which the authors could not explain at that stage.

Kline and Sauer[23] also reported the existence of low temperature peaks, whereas Kaelble[24] studying the same resin cured with a long-chain aliphatic diamine or a chlorosubstituted DDM could not detect the low temperature peak at temperatures down to $-50\,°C$.

Other studies on dynamic mechanical properties have been carried out by Jenkins[25] and Jenkins and Karre[26] using a vibrating reed technique. In the latter work the polymerisation of an epoxide resin with DTA or MPD as curing agents was studied using both infra-red spectroscopy and dynamic mechanical testing, and kinetic data of the reaction were obtained (Section 3.8). Jenkins and Karre *(op. cit.)* also remarked on the effect of freezing in the polymer segmental motion as T_g advanced beyond the cure temperature. In the glassy state the rate of segmental motion is extremely low, and thus available reactive sites on the polymer chain do not readily take up positions for reaction. When the cure temperature is equal to the glass transition temperature, reaction can occur more readily; when it exceeds T_g, cross-linking continues until T_g is equal to or greater than the cure temperature.

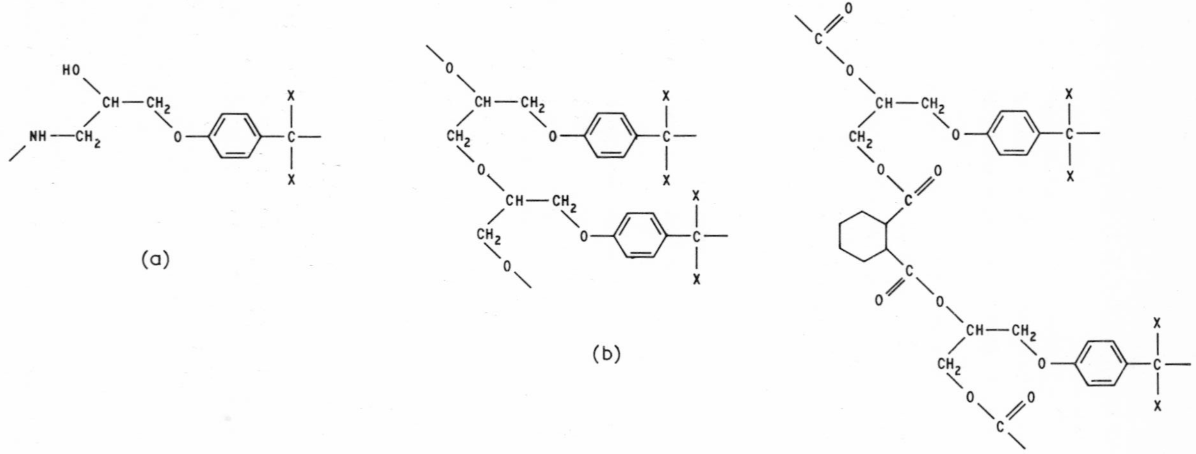

Fig. 5.3 (a) Amine cure (gives low temperature peak); (b) boron trifluoride cure (no low temperature peak); (c) anhydride cure (no low temperature peak)

X = H, Me or Ph

A torsion pendulum was used by Kwei[27] to study the pure diglycidyl ether of DPP when cured with a series of long-chain diamines. Measuring T_g via dilatometry and damping curves, Kwei correlated the values obtained with the number of methylene groups between cross-link junctions in the aliphatic amine part of the network. The greater the number of methylene groups between amino groups, the lower was observed T_g. In an interesting study, Galperin[28] used the torsion pendulum to examine a resin cured with hexamethylenediamine and filled with 0–40% titanium dioxide, over the temperature range 20–90 °C. He observed that increasing filler content caused significant immobilisation of cross-linked polymer chains, and hence greater stiffness and a consequent increase in T_g accompanied by a decrease in damping capacity (see, however, Section 5.3.4, ref. 11).

5.3.6 ELECTRICAL VOLUME RESISTIVITY

Whilst using the change in electrical volume resistivity with time at various temperatures to determine rates of cure and energies of activation (Section 3.8), Warfield and Petree[8] suggested that the temperature-dependence of resistivity of the system could be related to the extent of polymerisation. They based this proposal on the observation that uncured resins had but a small temperature-dependence of resistivity, whereas cured resins had a high dependence.

5.3.7 HARDNESS: DEGREE OF SWELLING

The extent to which a cured epoxide resin swells by absorption of a solvent has been used[3, 29] as a measure of the cross-linking of the system. Similarly the softening point of the cured resin as determined by the extent of penetration of a needle at different temperatures has also been used as a measure of degree of cure.[3]

Vicat's method is one of the most widely used penetration tests, and employs a flat-ended needle of 1 mm diam. cross-section, which is forced under a load into the material. The temperature is raised at 50 degC per hour, the temperature at which the needle has sunk 1 mm being called the Vicat temperature.

5.4 MORPHOLOGY OF CURED EPOXIDE RESINS

So far in this Chapter the structure of the cured resin has been considered at the molecular level, but not the way in which the cross-linked and non-cross-linked molecules are distributed throughout the polymer. There are two possibilities—first, that the cross-linked network is one giant molecule rather like a sponge, the uncured material occupying the holes in the sponge-like network; secondly, that the cross-linked molecules may be aggregated together in well defined regions, surrounded by or embedded in the unpolymerised molecules. Evidence has been obtained by Erath *et al.*[30,31] indicating that the latter alternative is the correct one. An electron microscopic study of cured resins including phenolics, silicones, DAP, and epoxides showed that in all but the silicones, globular formations or micelles existed in the cured polymer, having diameters of 400–900 A. Thus the cured resin molecules were relatively small three-dimensional networks whose ultimate size was determined by limiting of intermolecular reaction by steric hindrance. Most of the micrographs showed the micelles in linear arrays, suggesting that chain growth had its inception along the resin filaments. These micelles probably influence the physical properties of the material markedly, and could explain the difference between theoretical and experimental values for tensile strength.

This observed micellar structure for cured epoxide resin has been confirmed by Cuthrell.[32-34] Several epoxide resin systems, including a diglycidyl ether of DPP and a cycloaliphatic resin, together with curing agents MPD and HPA, were found to be two-phase systems possessing roughly spherical floccules arranged in layers in an interstitial fluid resembling the starting material. The size of the floccules apparently depended upon the initial cure rate, and properties such as density, hardness, glass transition point, and dielectric strength were all related to this size. Moreover, the surface properties of the system were found to be different from those of the bulk material. Cuthrell showed that the mould in which the polymers were cured produced a material gradient in the polymer surface region similar to the electrical double layer at phase boundaries in colloidal systems.

5.5 PHYSICAL PROPERTIES

The large number available of different resins, curing agents, diluents, etc., makes possible the preparation of cured epoxides with widely differing properties. They can be made highly flexible and rubber-like, or hard and brittle. They can have high or low heat deflection temperatures, and can if necessary be made flame-retardant. However, in general, an unfilled casting resin would have physical properties within the broad limits of Table 5.2.

Table 5.2 SOME GENERAL PHYSICAL PROPERTIES OF AN UNFILLED CASTING EPOXIDE RESIN

Specific gravity	1·2–1·3
Hardness (Rockwell M scale)	100–110
Ultimate tensile strength, lb/in^2	4–13 000
Elongation at break, %	3–5
Young's modulus, lb/in^2	2–5 × 10^5
Ultimate compressive strength, lb/in^2	15–30 000
Impact strength (notched Izod); ft–lb/in notch	0·3–0·9
Thermal conductivity (cal cm^{-1}sec^{-1}degC^{-1})	4–5 × 10^{-4}

The published information on the strength properties of epoxides is limited to ultimate values; yet the engineer requires more comprehensive design data such as the full stress/strain curve, perhaps at a number of different temperatures and at various rates of testing. Unfortunately data of this nature are not readily available and are not seen in the suppliers' technical literature.

The thermal properties of interest, in addition to thermal expansion and conductivity, are short- and long-term heat stability and fire resistance. The flammability of cured epoxides is frequently determined by the ASTM D635–56 Test. This classifies materials as burning, non-burning, or self-extinguishing, and defines burning rate in inches per minute, but does not specify a maximum elapsed time for a resin to extinguish itself. The current practice is to regard 15 sec as the maximum time for satisfactory self-extinguishing properties. Normally all cast epoxide systems will burn, but can be made self-extinguishing by incorporating halogenated resins or curing agents. Glycidyl ether resins based on a brominated DPP are available commercially, containing 18–20% or 45–48% bromine. When used as a blend with the normal non-halogenated resin, these give self-extinguishing properties, flame-retardancy being achieved by the bromine liberated at the decomposition temperature.

Other less popular ways of improving flame-retardancy in resins are to use HET anhydride as the curing agent, to incorporate quantities of antimony trioxide as a filler, or to use a non-reactive phosphorous-containing diluent. All of these methods either lead to deterioration in the physical properties of the system or limit the choice of curing agent and fillers normally available.

The heat resistance or thermal stability of cured epoxides has become of increasing importance over the last few years, reflecting the growing use of epoxide resins in rockets and high-performance aircraft. This has led numerous workers to investigate and compare the thermal stability of many different resin combinations and to attempt to identify the mechanism of thermal degradation. Heat-ageing studies seeking to compare various systems are often carried out by the prolonged heating of samples at a fixed temperature in an oxidising atmosphere, the change in weight caused by this treatment being taken as a measure of thermal stability. A general finding has been that resins cured with acid anhydride have greater thermal stability than amine-cured resins, and that epoxidised novolaks are more stable than the diglycidyl ethers of DPP.[35] The greater stability of films cured with phthalic anhydride as compared with aromatic and aliphatic amine cured films was also confirmed by Conley,[36] who studied the oxidative degradation of thin films by means of their infra-red spectra.

More fundamental studies of the degradation process have been carried out using techniques such as vacuum and hot-wire pyrolysis, thermogravimetric analysis (TGA), and differential thermal analysis (DTA). Probably the most direct way of investigating thermal degradation is to heat the sample in vacuo and examine the products evolved, by using standard techniques such as separation of products by gas chromatography, and their identification by infra-red or mass spectrometry or by classical analytical procedures. Alternatively small samples of the polymer can be pyrolysed on an electrically heated filament, the volatile products formed being swept into a gas chromatograph by the carrier-gas stream, usually nitrogen or helium.

The technique of DTA, already briefly described (Section 5.3.4) has been used to determine the temperatures at which exothermic and endothermic reactions occur in the sample. It is, however, of limited value in determining the mechanism of thermal degradation, and in some cases DTA cannot distinguish between an exothermic reaction and an endothermic one which it is masking in the same temperature range.[37] TGA is the measurement of the weight loss of a sample as it is heated at a steady rate, often 1 degC/min, through a given temperature range. It is therefore not subject to the complications that can occur with DTA, and can be used to give a more reliable indication of the thermal stability of polymers, together with the kinetics and overall activation energy of the degradation reactions.

Vacuum pyrolysis has been used by Sugito and Ito,[38] Madorsky and Straus,[39] and Niemann *et al.*[40,41] and hot-wire pyrolysis by

Smith and co-workers.[42,43] Both techniques were used by Lee[44] in addition to DTA and TGA, in a very comprehensive study of glycidyl ether resins alone and polymerised with NMA and DDM curing agents. The uncured resins showed an exothermic peak on the DTA thermogram at 300–380°C, similar to that found by Anderson[45,46] and thought to be due to the isomerisation of the epoxide groups of the glycidyl ethers to aldehydes, and to some thermal homopolymerisation of the resins. The homopolymerisation of glycidyl ethers of DPP by heating for 10 hr at 200°C was earlier reported by Golubenkova.[47]

Anderson[37,45,46,48,49] has described the use of DTA and TGA techniques to investigate the thermal degradation of uncured and cured epoxides, and has used information collected in this way[49] to calculate an average activation energy for the degradation reaction amounting to 31–34 kcal/mole. Niemann[52] also found a similar value. TGA has also been used by Torossian and Jones[50] and Fleming,[51] who identified two degradative mechanisms for NMA-cured resins (see also Anderson[48]).

The work published so far has established that uncured glycidyl ether resins exhibit an exothermic change in the temperature range 300–400°C, attributed to isomerisation of epoxide to aldehyde groups, homopolymerisation of the resin, and some volatilisation and degradation.[44,45] Cycloaliphatic resins, where the epoxide group is internal and different in properties from the terminal (glycidyl ether) epoxide groups, have not been found to show this effect. The decomposition of the cured glycidyl ethers in an oxidising atmosphere can commence at temperatures as low as 150°C when an amine curing agent is used, but anhydride-polymerised systems are stable up to 200°C.[52] Castings of glycidyl ethers including novolaks cured with aniline-formaldehyde (AF) resins, HPA, and BF_3–MEA complex, have been kept in air at 200°C for 1000 hours.[35] The novolak–HPA system lost only 2% of its weight, but 6·4% was lost when BF_3–MEA was used as the curing agent. The solid DPP-ECH resins showed weight losses of 4·8% with the AF resins, 6·3% with HPA, and 5·7% with BF_3–MEA. The liquid DPP–ECH resin gave comparable figures of 9·1%, 3·7%, and 12·4%. In contrast, other workers[50] found that the short-term weight loss of different epoxide systems (i.e., DPP–ECH resin cured with DTA and DDM and an epoxidised novolak with BF_3–MEA, when heated in air at 175°C for 16 hr) was extremely small, only about 0·1%.[50] In a comprehensive study, Lee[44] found that an-hydride-cured glycidyl ether resins began to lose weight at

260–268°C when heated at 9–10 degC/min in vacuo.

The actual mechanism of thermal degradation is not clear, and much remains to be done, as is made clear in a valuable review by Bishop and Smith.[53]

5.6 ELECTRICAL PROPERTIES[54–56]

Epoxide resins are widely used for electrical insulation. They have very good electrical properties and in addition show outstanding adhesion to many substrates, low shrinkage during and after cure, and can be easily formed into complex shapes by casting techniques. Moreover their mechanical, physical, and chemical properties are extremely good.

Electrical properties which are important in the choice of insulating materials are:

(*a*) volume resistivity,

(*b*) surface resistivity,

(*c*) dielectric strength,

(*d*) power factor, and

(*e*) permittivity (dielectric constant).

These properties are not usually constant, but vary with time, frequency of applied voltage, and temperature of the material.

The extent to which a material can withstand the passage of a current is given by its volume and surface resistivities, which are usually measured in ohm cm and ohms respectively. A typical value for volume resistivity of epoxide systems is $> 10^{16}$ ohm cm.

Breakdown in cast insulation can be caused by discharges within the insulation if air voids or stress cracks are present because of incorrect manufacture, and surface tracking (see Section 9.3.2). A further way in which insulators can fail is by thermal breakdown. Apart from the usual heating effects caused by a direct current passing through a material, the passage of an alternating current can also cause heat build-up by dielectric heating. In the latter situation the energy loss in a dielectric is proportional to the square of the applied voltage, its frequency and $K\tan\delta$ (i.e., dielectric constant multiplied by the loss factor). Hence in power applications where frequencies of 50 Hz are met, very low values of dielectric constant and loss factor and high values of resistivity are needed, especially at high temperatures, to avoid thermal breakdown due to 'run away' dielectric heating. Epoxide resins have typical dielectric constants at 25 °C and 10^3 Hz of about 3·0–4·5 and power factors ·

of 0·004–0·040, both of these properties increasing with increasing temperature, the change being more marked at lower frequencies. Fig. 5.4 illustrates the variation in these properties with frequency and temperature for one casting system.

In general for electronic applications where very high frequencies are used, e.g., 20 MHz, the need is for constancy in electrical

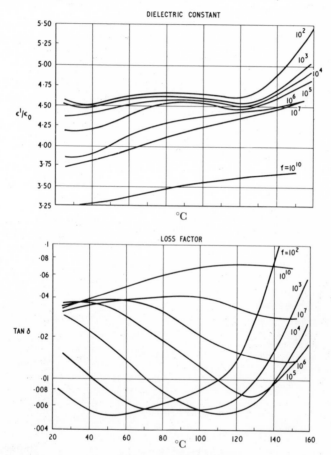

Fig. 5.4 Resin system: liquid diglycidyl ether resin cured with 14 phr curing agent MPD. Cured 2 hr at 85°C plus 4 hr at 150°C[57]

properties with time and in varying environments.

A most comprehensive consideration of the electrical and

mechanical properties of epoxide resins and their variation due to factors such as frequency and temperature is given by Clark.[56]

5.7 CHEMICAL PROPERTIES

The outstanding chemical resistance of a properly formulated and well-cured epoxide resin system is widely known and is used in applications such as monolithic flooring, tank linings, and surface coatings. This high level of resistance depends upon achieving maximum cross-linking in the cured polymer, and also upon the nature of these linkages. It is therefore possible to make some general predictions of the chemical resistance of various systems.

Ether links, developed through resin homopolymerisation with catalytic curing agents, would be expected to be stable against most inorganic and organic acids and alkalis. Ester links, which are present in anhydride-cured systems would not however be expected to be highly resistant to strong alkalis and inorganic acids. The C–N

Table 5.3 CHEMICAL RESISTANCE OF CASTINGS CURED WITH MPD: IMMERSED 180 DAYS[57]

Chemical	Change in appearance of specimen	Change in flexural strength lb/in^2	
		Original	After immersion
Immersion at 54°C			
25% hydrochloric acid	Slightly greenish	19 800	16 300
25% acetic acid	No change	—	18 400
100% trichloroethylene	No change	—	15 300
6% sodium hypochlorite	Heavy chalking	—	19 700
Distilled water	No change	—	18 600
Immersion at 82°C			
50% sodium hydroxide	Slight surface dullness	—	20 900
25% sulphuric acid	Slight surface dullness	—	15 000
25% hydrochloric acid	Considerable darkening	—	15 600
40% formaldehyde	Slight swelling	—	14 400
25% chromic acid	Slight chalking	—	18 300

bond, formed during amine polymerisation of epoxides would be expected to show stability towards inorganic alkalis and acids, but not towards organic acids.

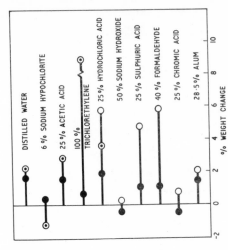

Fig. 5.6 Percentage weight change of unfilled epoxide resin/curing agent MPD castings after 180 day immersion. 15 phr curing agent MPD; cured 2 hr at 85°C plus 4 hr at 150°C[57]

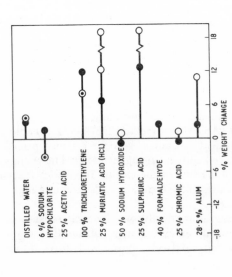

Fig. 5.5 Percentage weight change of unfilled epoxide resin/curing agent TET castings after 180 day immersion. 15 phr curing agent TET; cured 21 days at 25°C[57]

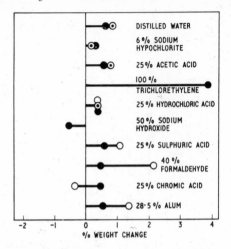

Fig. 5.7 Percentage weight change of unfilled epoxide resin/phthalic anhydride castings after 180 day immersion. 75 phr curing agent phthalic anhydride plus 0·1 phr benzyldimethylamine; cured 3 hr at 120°C plus 4 hr at 150°C[57]

A measure of the chemical resistance of a casting can be obtained by the weight change or variation in mechanical properties after immersion in a reagent (ASTM test method D534–52T). Table 5.3 gives some typical chemical resistance data for castings prepared from an epoxide resin cured with curing agent MPD. The change in flexural strength after immersion of the casting in various reagents for 180 days at temperatures of 54°C and 82°C is used as a measure of the chemical resistance.

Figs. 5.5–5.7 show diagrammatically the chemical resistance, as measured by their percent weight change, of unfilled castings prepared from a liquid diglycidyl ether cured with three different curing agents.

REFERENCES

1. MIKA, T. F., Private Communication
2. WARFIELD, R. W. and PETREE, M. C., *Makromolek. Chem.*, **58,** 139 (1962)
3. DANNENBERG, H. and HARP, W. R., *Analyt. Chem.*, **28,** 86 (1956)
4. BURHANS, A. S., PITT, C. F., SELLARS, R. F. and SMITH, S. G., 21st meeting of Reinforced Plastics Division of SPI, Chicago (1965)
5. GORDON, M., *High Polymers*, published for The Plastics Institute by Iliffe, London (1963)
6. DANNENBERG, H., *S.P.E. Trans.*, **3,** 78 (1963)
7. DAMUSIS, A., *Am. chem. Soc. Symp.* (1956)
8. WARFIELD, R. W. and PETREE, M. C., *S.P.E. Trans.*, **1,** 3 (1961)
9. HARROD, J., *J. Polym. Sci.*, Part A, **1,** 385 (1963)
10. HOERNER, A. H., COHEN, M. and KOHN, L. S., *Mod. Plast.*, **35,** No. 1, 184 (1957)
11. KWEI, T. K. and KUMMINS, C. A., *J. appl. Polym. Sci.*, **8,** 1483 (1964)
12. KLOPFENSTEIN, E. and LEE, H., *Insulation, Lake Forest*, **4,** No. 2, 13 (1958)
13. WARFIELD, R. W., *S.P.E. Jl.*, **15,** 625 (1959)
14. MURPHY, C. B., *Mod. Plast.*, **37,** No. 12, 125 (1960)
15. ANON., *Applied Polym. Symp., Interscience*, **2** (1966)
16. DOUBLE, J. S., *Trans. Plast. Inst. Lond.*, **34,** No. 10, 73 (1966)
17. FAVA, R. A., *Polymer*, **9,** No. 3, 137 (1968)
18. DRUMM, M. F., DODGE, C. W. H. and NIELSEN, L. E., *Ind. Engng. Chem.*, **48,** 76 (1956)
19. NIELSEN, L. E., *Mechanical Properties of Polymers*, Reinhold, New York, Chap. 7 (1962)
20. KAELBLE, D. H., *S.P.E. Jl.*, **15,** 1071 (1959)
21. KLINE, D. E., *J. Polym. Sci.*, **47,** 237 (1960)
22. MAY, C. A. and WEIR, F. E., *S.P.E. Trans.*, **2,** No. 3, 201 (1962)
23. KLINE, D. E. and SAUER, J. A., *S.P.E. Trans*, **2,** No. 3, 21 (1962)
24. KAELBLE, D. H., *J. appl. Polym. Sci.*, **9,** 1213 (1965)
25. JENKINS, R. K., *J. appl. Polym. Sci.*, **11,** 171 (1967)
26. JENKINS, R. and KARRE, L., *J. appl. Polym. Sci.*, **10,** 303 (1966)
27. KWEI, T. K., *J. Polym. Sci.*, **A-2,** No. 11, 943 (1966)
28. GALPERIN, I., *J. appl. Polym. Sci.*, **11,** 1475 (1967)
29. KWEI, T. K., *J. Polym. Sci.*, **A-1,** 2977 (1963)
30. ERATH, E. H. and SPUR, R. A., *J. Polym. Sci.*, **35,** 391 (1959)
31. ERATH, E. H. and ROBINSON, M., *J. Polym. Sci.*, **C,** *Polym. Symp.*, No. 3 (1963)
32. CUTHRELL, R. E., *J. appl. Polym. Sci.*, **11,** 949 (1967)
33. CUTHRELL, R. E., *J. appl. Polym. Sci.*, **11,** 1495 (1967)
34. CUTHRELL, R. E., *J. appl. Polym. Sci.*, **12,** 1263 (1968)
35. LEMON, P. H. R. B., *Br. Plast.*, **36,** 336 (1963)
36. CONLEY, R. T., *S.P.E. Ann. tech. Conf. Proc.*, 118 (1964)
37. ANDERSON, H. C., *Nature*, **191,** No. 4793, 1088 (1961)
38. SUGITO, T. and ITO, M., *J. chem. Soc., Japan*, **38,** 1670 (1965)
39. MADORSKY, S. L. and STRAUS, S., *Mod. Plast.*, **38,** No. 6, 134 (1961)
40. NIEMANN, M. B., GOLUBENKOVA, L. I., KOVARSKAYA, B. M., STRIZHKOVA, A. S., LEVANTOVSKAYA, I. I., AKUTIN, M. S. and MOISEEV, V. D., *Vysokomolek. Soedin*, **1,** 1531 (1959)
41. NIEMANN, M. B., KOVARSKAYA, B. M., GOLUBENKOVA, L. I., STRIZHKOVA, A. S., LEVANTOVSKAYA, I. I. and AKUTIN, M. S., *J. Polym. Sci.*, **56,** 383 (1962)
42. STUART, J. M. and SMITH, D. A., *J. appl. Polym. Sci.*, **9,** 2195 (1965)
43. KEENAN, M. A. and SMITH, D. A., *J. appl. Polym. Sci.*, **11,** 1009 (1967)

44. LEE, L-H., *J. Polym. Sci.*, Part A, **3,** 859 (1965)
45. ANDERSON, H. C., *Analyt. Chem.*, **32,** 1592 (1960)
46. ANDERSON, H. C., *Polymer*, **2,** 451 (1961)
47. GOLUBENKOVA, L. I., KOVARSKAYA, B. M., AKUTIN, M. S. and SLONIMSKII, G. L., *Kolloid Zh.*, **20,** 29 (1958)
48. ANDERSON, H. C., *J. appl. Polym. Sci.*, **6,** 484 (1962)
49. ANDERSON, H. C., *Kolloidzeitschrift*, **184,** No. 1, 26 (1962)
50. TOROSSIAN, K. A. and JONES, S. L., *J. appl. Polym. Sci.*, **8,** 489 (1964)
51. FLEMING, G. J., *J. appl. Polym. Sci.*, **10,** 1813 (1966)
52. NIEMANN, M. B., KOVARSKAYA, B. M., YAZVIKOVA, M. P., SIDNEV, A. I. and AKUTIN, A. S., *Vysokomolek, Soedin*, **3,** 602 (1961)
53. BISHOP, D. P. and SMITH, D. A., *Ind. Engng. Chem.*, **59,** No. 8, 32 (1967)
54. JACKSON, W., *The Insulation of Electrical Equipment*, Chapman & Hall, London (1951)
55. MASON, J. H., *Trans. Plast. Inst. Lond.*, **30,** 87, 171 (1962)
56. CLARK, F. M., *Insulating Materials for Design and Engineering Practice*, Wiley, New York (1962)
57. Shell Chemical Co., Technical Literature

CHAPTER 6

Modifying Materials

The large number of available epoxide resins and curing agents makes it possible to obtain castings with a wide spectrum of properties. Rigid or flexible castings can be combined with outstanding adhesion, mechanical strength, and electrical insulation, coupled with stability in widely differing environments.

This range of behaviour can be extended even further by the incorporation of diluents, fillers, extenders, flexibilisers, plasticisers, etc., into the basic resin mixture before curing. Epoxide resins are also used with fibrous reinforcements to produce composites, such as glass cloth laminates. This Chapter considers the main types of modifying materials but does not discuss composite materials, which are dealt with in Chapter 10.

6.1 DILUENTS

In many applications low viscosity resins are desirable. They are easy to handle and mix with other materials. They also allow more filler to be incorporated with the resin and give better impregnation of glass cloth. For these reasons diluents are often used in epoxide resin formulations although other properties of the system are also modified in consequence. Pot life, exotherm, HDT, physical, and electrical properties may all be affected. Some compounds can be used deliberately to modify two separate characteristics, e.g. viscosity and flexibility, or viscosity and cost. Compounds considered here are those whose prime use is for viscosity reduction.

6.1.1 NON-REACTIVE DILUENTS

Aromatic hydrocarbons such as toluene or xylene can bring about significant reductions in the viscosity of the low-M resins, 5 phr of xylene reducing the viscosity from 120 to about 20 poises. But this reduction is accompanied by a 15–25% decrease in compressive yield strength and 10–20% reduction in compressive modulus. The solvent also remains in the cured system but is not chemically bound. Thus the casting will have inferior chemical resistance and if it is heat-cured the diluent can volatilise and cause blow-holes or bubbles.

A popular non-reactive diluent is dibutyl phthalate, used at a concentration of 15–17 phr with a liquid resin which gives a viscosity of 20–25 poises. This phthalate reduces exotherm during cure as well as having an external plasticising effect on the cured resin. The physical properties of the plasticised resin when cured with TET are given in Table 6.1.

Table 6.1 TYPICAL PROPERTIES OF CASTINGS PREPARED WITH A PLASTICISED RESIN[*1]

Pot life (500 g; 23 °C)	55 min
Peak exotherm temperature (500 g)	182 °C
Time to reach peak exotherm	70 min
Cure schedule	30–60 min at 100 °C
Heat deflection temperature, °C	
(ASTM D648–56)	52
Tensile strength, lb/in^2 (ASTM D638–60T)	7 100
Volume resistivity, ohm cm	
(ASTM D257–58)	7×10^{16}
Power factor; 1 KHz (ASTM D257–58)	0·016
Dielectric constant (ASTM D150–59T)	
1 KHz	4·25
10 KHz	4·19

*Resin: diglycidyl ether of DPP (WPE 182–194)
Plasticiser: dibutyl phthalate (15–17 phr)
Curing agent: TET (10 phr)

Pine oil may also be used as a diluent at about 20 phr without a drastic reduction in overall properties.

6.1.2 REACTIVE DILUENTS

This class of substance usually though not always contains epoxide groups which join in the polymerisation reactions and become chemically bound into the network. Whenever possible the diluent should effect a considerable reduction in viscosity when used at low concentrations. In addition it should be non-reactive with the resin under normal storage conditions and react with a curing agent at about the same rate as the resin.

In practice, most reactive diluents are monoepoxide compounds and hence reduce the cross-link density of the system. This is shown by the lower HDTs and general heat resistance of diluted systems. Some polyepoxides, however, which are more properly regarded as epoxide resins in their own right, can improve properties, an example being the use of vinylcyclohex-3-ene dioxide (VCDO) in a system where its epoxide groups are able to become fully reacted. Other cycloaliphatic resins, because of their low viscosity, have also been used with the liquid diglycidyl ether resins or with the solid epoxidised novolak resins, often with much less deterioration in high-temperature properties than would be obtained with a mono-functional diluent, and sometimes even with the enhancement of properties.

Some of the commercially more important mono- and di-epoxide diluents are shown in Fig. 6.1; and Table 6.2 shows their effect on certain physical properties of room and elevated temperature

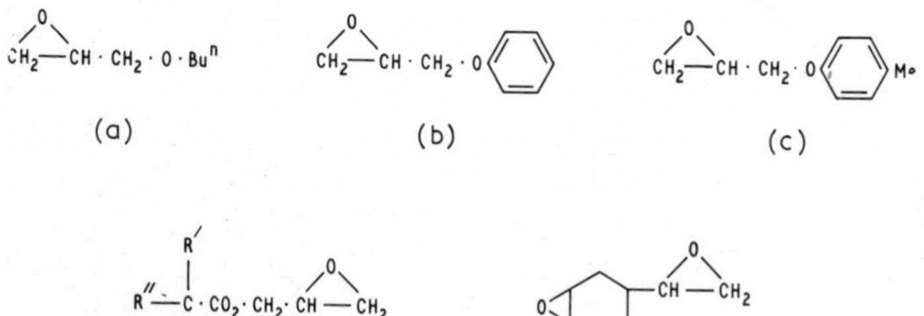

Fig. 6.1 Some mono- and di-epoxide reactive diluents: (a) n-butyl glycidyl ether (n-BGE); (b) phenyl glycidyl ether (PGE); (c) p-tolyl ('cresyl') glycidyl ether (CGE); (d) glycidyl ester of a tertiary carboxylic acid (Cardura* E); vinylcyclohex-3-ene dioxide (VCDO); (f) tetramethylene diglycidyl ether (*Shell Chemical Co. Ltd.)

Table 6.2 EFFECT OF DILUENTS ON CURED PROPERTIES[1]

Diluent	None	BGE(a)	PGE(b)	CGE(c)	VCDO(d)	γ-BL(e)	CARDURA E(f)
	A. Room temperature cures (7 days at room temperature)						
Concentration of diluent, %w	—	12·0	20·0	24·0	23·5	13·5	22·5
DTA, phr	12·0	11·5	11·4	11·4	15·0	9·5	10·5
Cure rate at 25°C	—	50	48	54	38	58	37
Barcol hardness	27	24	26	25	25	22	18
Tensile properties (rate of crosshead travel—0·5 in/min)							
Tensile strength at break lb/in² ($\times 10^{-3}$)	5·8	9·2	9·5	9·3	8·4	8·1	9·2
Tensile modulus lb/in² ($\times 10^{-5}$)	4·5	4·6	5·4	5·2	4·8	3·7	4·7
Elongation at break, %	1·4	3·4	2·0	2·0	3·0	2·9	3·3
Compressive properties (rate of crosshead travel—0·05 in/min)							
Compressive strength at failure lb/in² ($\times 10^{-3}$)	10·5	14·8	15·6	16·5	11·8	12·6	14·0
Compressive modulus, lb/in² ($\times 10^{-5}$)	3·5	4·5	5·3	5·5	3·9	4·6	4·1
Deformation at failure, %	3·8	4·6	4·4	4·1	4·1	4·0	4·6

(a) = Butyl glycidyl ether
(b) = Phenyl glycidyl ether
(c) = Tolyl glycidyl ether
(d) = Vinylcyclohexene-3-ene dioxide
(e) = γ-Butyrolactone
(f) = CARDURA E (Shell Chemical Company's glycidyl ester)

Table 6.2 (*continued*) EFFECT OF DILUENTS ON CURED PROPERTIES[1]

Diluent	BGE[a]	PGE[b]	CGE[c]	VCDO[d]	γ-BL[e]	CARDURA E[f]
B. Elevated temperature cures (2 hr at 80°C plus 2 hr at 150°C)						
Concentration of diluent, %w	12·0	20·0	24·0	23·5	13·5	22·5
Aromatic diamine, phr*	22·0	22·0	22·0	28·0	18·0	20·0
Cure rate at 65°C	81	64	60	63	45	81
HDT, °C	109	104	96	135	85	87
Barcol hardness	31	34	31	41	28	25
Tensile properties (rate of crosshead travel— 0·05 in/min)						
Tensile strength at break, lb/in² ($\times 10^{-3}$)	15·0	10·3	11·9	17·0	9·2	12·0
Tensile modulus, lb/in² ($\times 10^{-5}$)	5·6	5·6	5·3	6·1	4·2	4·1
Elongation at break, %	5·8	2·1	4·0	5·6	4·6	5·6
Compressive properties (rate of crosshead travel— 0·05 in/min)						
Compressive strength at failure, lb/in² ($\times 10^{-3}$)	16·1	16·4	18·0	22·5	12·9	16·4
Compressive modulus, lb/in² ($\times 10^{-5}$)	4·4	4·6	5·6	4·8	4·0	4·8
Deformation at failure, %	7·4	6·1	5·3	9·0	7·3	6·0

* Eutectic mixture of aromatic diamines

(a) = Butyl glycidyl ether
(b) = Phenyl glycidyl ether
(c) = Tolyl glycidyl ether
(d) = Vinylcyclohex-3-ene dioxide
(e) = γ-Butyrolactone
(f) = CARDURA E (Shell Chemical Company's glycidyl ester)

curing systems. Fig. 6.2 illustrates the viscosity-reducing effect of different concentrations of diluents.

Care must be exercised when handling some of the reactive monoepoxide compounds since they are more prone to causing dermatitis than many of the resins themselves. Moreover some of the diepoxides such as VCDO are also not recommended for general

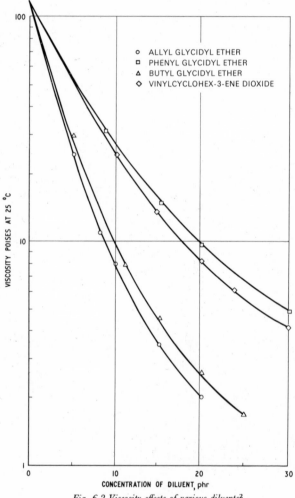

Fig. 6.2 *Viscosity effects of various diluents*[2]

use as a diluent because of their possible harmful physiological effects.

Two commercial reactive diluents not containing epoxide groups are triphenyl phosphite and γ-butyrolactone.

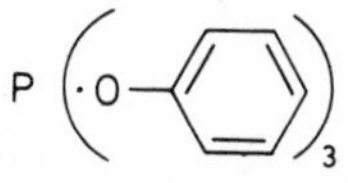

Triphenyl phosphite 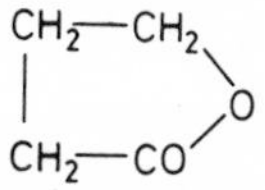　　　　γ-butyrolactone

The phosphite is a low-viscosity colourless liquid, which is sensitive to moisture and reacts with hydroxyl groups in the resin forming an ester and liberating phenol (Fig. 6.3).

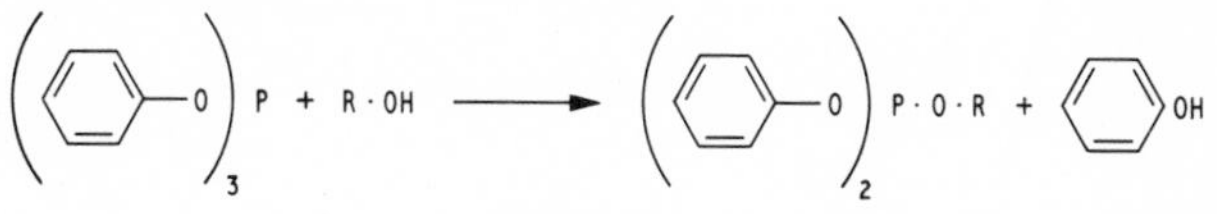

Fig. 6.3

The phenol may then react with the epoxide group of the resin or serve as an accelerator in the curing reaction. This accounts for the instability of mixtures of resin and phosphite, there being an increase in viscosity of the mixture under storage.

γ-Butyrolactone is an effective viscosity-reducer,[3] 10 phr reducing the viscosity of a liquid diglycidyl ether resin from about 150 to

Fig. 6.4

20–25 poises. In the curing reaction with amines, lactones react with formation of an amide, which can then cross-link with the polymer via the hydroxyl group formed (Fig. 6.4).

6.2 FILLERS

The changes caused in epoxide resin systems by the incorporation of fillers may be summarised as follows.[4]

Advantages of fillers	*Disadvantages of fillers*
Lower cost of product	Increased weight
Reduced shrinkage on curing	Increased viscosity

Decreased exothermic temperature rise on curing
Increased thermal conductivity
Reduced expansion and contraction with temperature change
Higher heat deflection temperatures
Improved heat-ageing properties
Reduced water absorption
Improved abrasion resistance
Improved toughness if fibrous fillers used
Increased surface hardness
Increased compressive strength
Increased electrical strength

Loss of transparency
Tendency to trap air
Difficulty of machining hard fillers
Decreased impact and tensile strengths with powdery fillers
Increased dielectric constant and power factor

Note that certain non-fibrous fillers have the reverse effect. These exceptions, usually not simple mineral powders, are represented in Table 6.3 by vermiculite and phenolic microballoons.

In practice most epoxide resin systems have fillers incorporated, the choice being determined by the property it is wished to improve, the composition of the resin, the curing agent, other components, and practical considerations. Clearly the type and amount of filler in an electrical application will differ from that in a trowelling mixture or in an adhesive formulation.

The great majority of the many commercial fillers available are inert materials which do not react with the components of the resin mixture. They can be organic or inorganic in nature, and spheroidal, granular, lamellar, or fibrous in shape (Table 6.3).

6.2.1 CHANGE IN VISCOSITY DUE TO FILLERS

All fillers increase the viscosity of the resin mixture, fibrous fillers having the greatest effect. Thus the maximum filler loading for any system is frequently set by the maximum viscosity allowable for its method of application. Table 6.4 shows the maximum amount of fillers tolerated in a mixture for pouring.[5]

The viscosity of a filled mixture is influenced by the viscosity of its individual liquid components and of the mixture to which they all contribute. The use of a diluent permits greater filler loading be-

Table 6.3 SOME TYPICAL PROPERTIES OF FILLERS

Name	Composition	Particle shape	Specific surface ($\frac{surface\ area}{volume}$)	Bulk density (lb/ft^3)	Characteristics and main use
Marble flour	Magnesium—Calcium carbonate	Granular	Medium	69–81	General purpose fillers. Used when castings to be machined
Chalk powder	Precipitated calcium carbonate	Crystalline	High	50–55	
Silica flour	Ground quartz	Granular	Medium	62–72	Standard filler for electrical castings. Difficult to machine
Mica flour	Muscovite (silicate)	Lamellar	High	19–24	Provides good crack-resistance against thermal and mechanical shock
Slate powder	Slate (silicates)	Mainly lamellar	Medium	43–55	General purpose filler. Good abrasion resistance
Zircon flour	Zircon (silicate)	Granular	Medium	105–120	Provides very high abrasion resistance
Sand	Quartz, feldspar, and other minerals	Spheroidal	Low	93–105	Cheap bulk filler giving high compressive strength
Vermiculite	Vermiculite (silicate)	Exfoliated laminae	High	6–9	Provides light weight bulking effect
Phenolic microballoons	Phenolic resin	Hollow spheres	Medium	6–9	
Aluminium powder	Metallic aluminium	Granular	Medium	62–69	Improves thermal conductivity
Asbestos fibre	Chrysotile (silicate)	Fibrous	Medium	24–31	Improves mechanical strength, especially impact strength
Chopped glass fibre	Low alkali glass	Fibrous	Medium	6–15	

cause of its overall viscosity-reducing effect. Raising the temperature of the mixture will also achieve the same effect although at the same

Table 6.4 MAXIMUM AMOUNT OF FILLER FOR A POURABLE MIXTURE[5]

Filler	Concentration (phr) at maximum pourable viscosity
Black iron oxide	300
Tabular aluminium oxide	200
Atomised aluminium	150
Graphite	50
Titanium dioxide	50
Calcium carbonate	30
Silica flour	30
Asbestos	5

time it shortens pot life. The particle size and shape of fillers also influences the increase in viscosity. As has been mentioned, fibrous fillers, and powders having a high surface area per unit weight, cause the maximum increase in viscosity.

Most fillers cause viscosity increases independent of the rate of shear applied to the system. However, a few special fillers such as colloidal silica, metallic leafing powders, asbestos floats, and kaolin impart a viscosity to the mixture which decreases at increasing rate of shear. This is a thixotropic effect, caused by the filler forming a temporary 'structure' in the mixture, which is broken down at high rates of shear. Thixotropic fillers are used to prevent run-off or sag in resin mixes that are applied to inclined or vertical surfaces.

6.2.2 EFFECT OF FILLERS ON EXOTHERM, POT LIFE, AND SHRINKAGE

Exothermic heat imposes a serious limitation to the bulk size of an epoxide resin casting. Resins are good thermal insulators, and the heat of the curing reaction is not easily dissipated. Extremely high temperatures can be obtained in the middle of castings, which can cause volatilisation of components of the resin system and extensive cracking and void formation.

For large castings, filled systems are therefore essential. Fillers assist by reducing the amount of resin used and by absorbing part of the heat evolved and hence reducing temperature high spots. In addition they increase the thermal conductivity of the system,[5, 6] although it remains extremely low compared with metals. Fibrous metal fillers give the best increase.[7] Table 6.5 gives some data on the thermal conductivity of a liquid resin system using different fillers.

Moderation of heat evolution also helps to extend pot life, particularly at high filler loadings. Shrinkage during cure is also reduced,

Table 6.5 THERMAL CONDUCTIVITY OF FILLED CASTINGS[1]

Filler	PHR	Wt. % of total	Vol. % of total	K (Btu/hr/ft²/ft/°F)
None*				0·13
Aluminium oxide	325	73	48	0·82
Atomised aluminium	200	63	45	0·94
Iron powder	300	72	30	0·52
Copper powder	1 200	91	60	0·94
Silica	150	56	39	0·44
			Steel =	26·0
			Aluminium =	116·0
			Copper =	224·0

*Cured with 20 phr eutectic mixture of aromatic amines.
  Cure schedule: 2 hr at 85°C, plus 2 hr at 150°C

the filler replacing part of the resin by an inert compound which does not undergo irreversible physical change on heating.

6.2.3 EFFECT OF FILLERS ON THE PHYSICAL PROPERTIES OF THE
       CURED SYSTEM

Fillers can bring about a number of changes in the physical properties of cured epoxides. In general, particulate fillers such as powdered metal oxides depress the ultimate tensile and compressive strengths (Table 6.6). However, they have little effect on impact

Table 6.6 EFFECT OF FILLERS ON ULTIMATE STRENGTHS OF CASTINGS*[1]

Filler	phr	Ultimate compressive strength (lb/in²)	Ultimate tensile strength (lb/in²)
None		28 000	12 500
Black iron oxide	400	18 000	4 500
Atomised aluminium	200	15 000	6 500
Asbestos floats	15	17 500	7 000
Water-ground mica	15	20 000	7 500

*Liquid diglycidyl ether resin + eutectic mixture of aromatic amines.
  Cure: 20 hr at 25°C, plus 24 hr at 65°C

strength, and increase compressive yield strengths as shown in Fig. 6.5. Aluminium powder is an exception, causing a decrease in compressive yield strength. Powdered fillers also tend to increase the surface hardness of castings, the effect varying with the nature

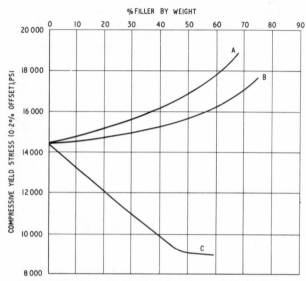

ALL CASTINGS BASED ON LIQUID DIGLYCIDYL ETHER RESIN PLUS 8 phr DIETHYLENETRIAMINE
STANDARD CURE OF 20 HR AT 25°C PLUS 2 HR AT 100°C USED IN ALL CASES

Fig. 6.5 Effect of filler concentration on compressive yield strength of castings.[1] Curve A = tabular aluminium oxide, curve B = black iron oxide, and curve C = atomised aluminium

of the filler. Zircon, slate powder, or silica flour can be used to increase abrasion resistance, which naturally is accompanied by an increased difficulty in machining compositions containing these powders.

Fibrous fillers such as asbestos, glass fibre, and nylon flock increase tensile and impact strengths, and small amounts of these fillers can be used with particulate fillers in applications where high impact strength is needed.

Inorganic fillers can also cause variations in the chemical and electrical properties of castings. In general most fillers, but especially fibrous ones, cause an increase in the water absorption of castings, and certain fillers diminish chemical resistance. Obvious examples are the effect of calcium carbonate on acid resistance, or of alumina on alkali resistance.

*Electrical components encapsulated in epoxide resin by Whiteley Electrical Radio Co. Ltd.
(By courtesy of CIBA (A.R.L.) Limited)*

*A section of an epoxide resin
encapsulated transformer by
Parmeko Ltd. showing iron
cores and primary and secondary
windings. On the left and right
the leads can be seen enclosed in
the encapsulation. (By courtesy
of CIBA (A.R.L.) Limited)*

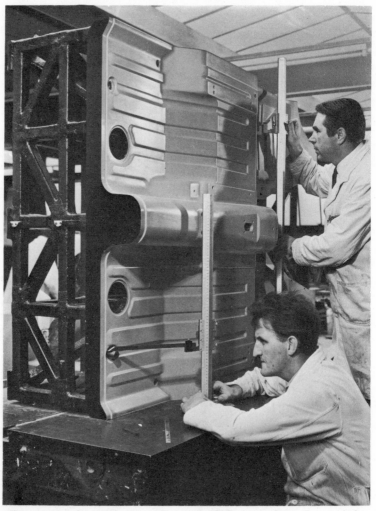

The construction of an epoxide resin negative master model for the manufacture of a car main floor-pan, by Technical Woodwork & Co. Ltd., Basildon, Essex. (By courtesy of CIBA (A.R.L.) Limited)

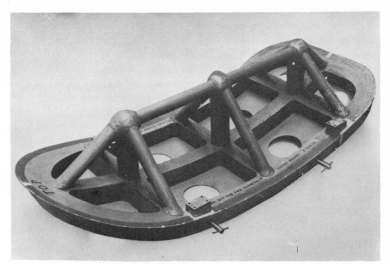

Male jig reference for the De Havilland DH 121 forward passenger entry door made in epoxide resins. (By courtesy of Shell Chemicals U.K. Limited)

Epoxide resin foundry pattern made by J. & E. Hall Ltd., Dartford, Kent, for casting the crankcases of Veebloc, high-speed refrigerating compressors. (By courtesy of CIBA (A.R.L.) Limited)

It is usual to employ non-conducting fillers in electrical casting systems and metallic fillers degrade electrical properties.[8] On the other hand, powdered copper or powdered flaked silver have been used to prepare electrically conducting castings.[9,10] Other approaches to produce conducting castings have been to use silver-coated copper spheres[11] or metal wool as conducting fibres.[12] The most popular electrical-grade filler is silica flour, which in use usually causes an increase in dielectric constant, arc resistance, and power factor, and a decrease in volume resistivity.

6.2.4 HANDLING FILLERS

All fillers should be dry, and those to be used for castings requiring good electrical properties should be specially dried before use. A typical heat treatment would be several hours at 110–130 °C, or 80 °C in vacuo. Once dried the fillers should either be used immediately or stored in airtight containers.

Apart from moisture, a number of other factors such as pH and particle size distribution can affect mix-viscosity and rate of cure. It is therefore important to eliminate batch-to-batch variation in fillers if reproducible characteristics are required. Many fillers are now given a surface treatment to improve the flow properties of the mix. One typical process is to coat the filler particles with calcium stearate.

In practice, for small mix-sizes it is usual to add the filler to the mixture of resin and curing agent. The curing agent frequently lowers the mix viscosity, which assists the incorporation of the filler and the escape of entrapped air. For large mixes, and especially in electrical work, it is preferred to mix filler with resin, the curing agent being added later. This procedure enables the resin and filler to be degassed in vacuo if necessary, and overcomes short pot life difficulties.

Thorough mixing is essential in all cases. On a small scale, hand mixing can be used and disposable containers are recommended. Larger scale operations require mechanical mixers with strong shearing actions, and the mixing vessel must sometimes be such as to allow a vacuum to be applied to the mix whilst the stirrer is in operation. Grinding machines such as three-roll mills can be used to achieve extremely thorough mixing. Epoxide resin castings that are unfilled, or filled with soft fillers such as chalk or marble flour, can be machined with ordinary metal working tools. When very hard fillers are used (e.g., silica, zircon flour, etc.), diamond-tipped tools are required.

6.3 RESINOUS MODIFIERS

A wide variety of resins can be mixed with epoxide resins not only to reduce cost but also in some instances to impart certain special properties to the resin system. Resins used primarily to achieve flexibility are discussed in Section 6.4, and those mainly used as co-reactants in surface-coating formulations in Chapter 8.

Coal tar pitch is the most widely used resinous modifier, mostly for surface-coating applications.[13] When used in high-solid coatings, films one coat thick can be applied having excellent water resistance. These paints are mainly used in marine, pipe, tank, and general industrial maintenance applications. When coal tar is incorporated into casting formulations it generally leads to a reduction in thermal and chemical resistance of the cured product, together with an increase in the flexibility and a decrease in tensile strength. Electrical insulation properties at room temperature are not greatly affected but are significantly lowered at elevated temperatures.

Petroleum-derived bitumens have also been used, especially in the U.S.A., as extenders for epoxide resin systems. One application involves the use of an epoxide resin/bitumen/dimer-trimer fatty acid combination as an overlay coating on aircraft runways and servicing areas. High-boiling petroleum distillates can also serve as low-cost extenders, but a compatibiliser such as an alkylphenol must be present in the mix to achieve compatibility between resin and extender.

Furfural resins can also be used with epoxides[14, 15] to reduce cost, and also to obtain an improvement in acid resistance.

Some other examples of the many resin blends examined are given below. In some, two resins may react with each other or may be bonded into one cross-link structure when the curing agent is added.

Epoxide-vinyl resin combinations are used not only in coating systems but also in adhesive formulations to improve peel-strength and impact resistance.[16] Epoxide resins are also used as stabilisers for PVC resins, usually in conjunction with a metallic salt. PVC has also been used as a flame-retardant filler in casting formulations.[17, 18]

Unsaturated polyesters used in the preparation of laminates and castings can co-react with those epoxide resins which also contain unsaturated centres, as in certain epoxidised olefins.

Finely divided PTFE has been added to epoxide resins in coating systems in an attempt to improve water sensitivity and flexibility.[19]

Many combinations of silicone resins with epoxides have been investigated mostly for surface-coating applications.[1]

6.4 FLEXIBILISERS

For certain applications it is desirable to modify the rather hard and brittle epoxide resin system in order to improve toughness or increase flexibility and thus improve the ability to withstand mechanical stressing or thermal shock. There are two ways in which such modification can be made to a cross-linked resin network:

(a) By introducing long flexible molecular chains, covalently linked to the network during curing (internal plasticisation). This is achieved by using flexibilising epoxide resins, curing agents, or reactive additives. In general, these compounds increase elongation and impact resistance at the expense of tensile strength. They also reduce other strength properties, electrical properties, chemical resistance, and HDT.

(b) By incorporating into the cured polymer long-chain molecules which remain unreacted with a cross-linked resin after cure (external plasticisation). This effect, frequently regarded solely as plasticisation and not flexibilisation, is achieved by the use of non-reactive or only partially reactive additives such as dibutyl phthalate.

6.4.1 FLEXIBLE EPOXIDE RESINS

The resins used are of two types, polyglycol diepoxides and epoxidised dimer fatty acids. The generalised structure of the polyglycol diepoxides is as follows:

$$CH_2-CH \cdot CH_2 \cdot O \cdot \left[CH_2 \cdot CH \cdot O \atop \overset{|}{R} \right]_n \cdot CH_2 \cdot CH \cdot O \cdot CH_2 \cdot CH-CH_2 \atop \overset{|}{R'}$$

R is usually a methyl group, as in a polypropylene glycol. By reacting the terminal hydroxyl groups of the glycol with ECH, a diglycidyl ether is easily obtained. At values of *n* not greater than 7,[20,21] these epoxides are low viscosity liquids, and are used in blends with other resins such as the liquid diglycidyl ethers or epoxidised novolaks in amounts up to 10–30% of the total resin. At this concentration, elongation, impact strength, and sometimes

Table 6.7 PROPERTIES OF A DIGLYCIDYL ETHER OF DPP RESIN AND AN EPOXIDISED NOVOLAK WITH VARYING PROPORTIONS OF A POLYGLYCOL DIEPOXIDE[*21]
(CURING AGENT USED THROUGHOUT: DDM)

	Ratio of glycidyl ether to polyglycol		Ratio of epoxidised novolak to polyglycol		
	100:0	70:30	100:0	90:10	70:30
Heat deflection temperature, °C	157	84	—	—	—
Flexural strength, lb/in²	16 970	14 060	11 000	13 400	16 900
Flexural modulus × 10⁻⁵	2·27	2·76	3·5	3·9	3·2
Compressive strength, lb/in²	16 970	12 160	33 000	31 200	31 800
Compressive modulus × 10⁻⁵	2·27	2·40	1·6	2·4	1·7
Tensile strength, lb/in²	6 930	9 160	7 200	6 800	5 400
Ultimate elongation, %	4·2	17·0	—	—	—
Izod impact strength, ft–lb/in notch	0·44	1·16	0·14	0·23	0·42

*The data for the glycidyl ether resin and the epoxidised novolak resin do not directly correlate because of differences in sample size, test method, and cure schedules.

tensile strength are increased, whereas flexural and compressive strengths are reduced, together with HDT. Chemical resistance may also be reduced and electrical properties are likely to fall off at higher temperatures. Table 6.7[21] gives data for one flexibilising resin of this type as a mixture with a liquid diglycidyl ether resin and an epoxidised novolak, both systems being cured with DDM.

Dimerised fatty acids have been treated with ECH to yield glycidyl fatty acid esters which are good flexibilising resins. A typical commercial product has a viscosity of 4–6 poises and WPE 400–420. Whilst it can be cured with most of the normal curing agents, it is normally used in conjunction with other resins similar to the epoxidised polyglycols. In general the epoxides based on dimer fatty acids increase the pot life of the conventional systems and longer cure schedules are required. Other flexibilising resins can be manufactured by reacting dimer fatty acids with a low molecular weight diglycidyl ether resin.

6.4.2 FLEXIBILISING CURING AGENTS AND REACTIVE ADDITIVES

These substances are similar to the flexible epoxide resins and usually consist of reactive groups which are separated by flexible molecular segments; they can therefore undergo cross-linking with the resin, becoming linked into the cured polymer network. Typical curing agents of this type are polyazelaic polyanhydride (PAPA), dodecenylsuccinic anhydride (DDSA), and the polyamides which have been discussed in Chapter 4.

$$HO \cdot \left[CO \cdot (CH_2)_7 \cdot CO_2 \right]_n \cdot H$$

$$Me \cdot (CH_2)_2 \cdot CHMe \cdot CH_2 \cdot CMe : CH \cdot CMe_2 \cdot CH \cdot CO$$
$$CH_2 CO \quad O$$

Poly(azelaic anhydride)   Dodecenylsuccinic anhydride (DDSA)
(Polyazelaic polyanhydride: PAPA)

Polysulphides of the following general formula $(n = 2-26)$ are also flexibilising additives.[22]

$$HS \cdot \left[(CH_2)_2 \cdot O \cdot CH_2 \cdot O \cdot (CH_2)_2 \cdot S \cdot S \right]_n \cdot (CH_2)_2 \cdot O \cdot CH_2 \cdot O \cdot (CH_2)_2 \cdot SH$$

The thiol-terminated polymers also occasionally have thiol groups along the molecular chain, the chains being cross-linked to an extent of about 2% of the total. The most commonly used polysulphide has a viscosity of 7–12 poises at room temperature and a molecular weight of about 1 000. Whilst they are able to react with epoxide resins via the thiol group, the polysulphides are always used in conjunction with a primary or tertiary amine curing agent at room or slightly elevated temperature. The higher temperatures that are required when aromatic amine and acid anhydride curing agents are used tend to degrade the polysulphide chains.

The polysulphides are usually used at concentrations between 25 and 100 phr which provide for fewer thiol groups than a 1:1 equivalence with epoxide groups. Moreover, the curing agent concentration is calculated from the epoxide content and the polysulphide ignored as a contributor to cross-linking. However, the reaction between epoxide and thiol groups leads to the formation of hydroxyl groups, which can then react with epoxide, especially under the influence of a tertiary amine. In addition, the amine curing agent will react in the usual way with the epoxide groups of the resin.

Some data on the mechanical properties of blends of epoxide with polysulphide resins are given in Table 6.8. Cranker and

Table 6.8 SOME MECHANICAL PROPERTIES OF POLYSULPHIDE–GLYCIDYL ETHER RESIN BLENDS[22]

	Three-component blend (parts by weight)			
Glycidyl ether resin	100	100	100	100
Polysulphide	100	100	50	50
Curing agent TET	10	—	20	—
Curing agent DMP–30	—	10	—	20
Viscosity (poise; 25 °C)	20	20	26	26
Pot life (50 g : min at 25 °C)	40	35	20	20
Tensile strength, lb/in^2	1 785	2 800	4 580	4 800
Elongation, %	30	30	50	10
Hardness (shore D)	41	63	60	78
After ageing for 70 hr at 100 °C				
Tensile strength, lb/in^2	1 800	900	5 000	4 200
Elongation, %	30	80	0	5
Hardness (shore D)	43	40	65	76

Breslau[22] also give further and more comprehensive data on the mechanical, electrical, and chemical performance of these blends.

REFERENCES

1. Shell Chemical Co., Technical Literature
2. MIKA, T. F., Private Communication
3. Ciba A.G., Belg. Pat. 624,986
4. Ciba (ARL) Ltd., Technical Literature
5. PARRY, H. L. and HEWITT, R. W., *Am. chem. Soc. Symp.* (1956)
6. PARRY, H. L. and HEWITT, R. W., *Ind. Engng. Chem.*, **49**, No. 7, 1103 (1957)
7. JURAS, A. W., U.S. Pat. 2,901,455
8. RINGWOOD, A. F., *S.P.E. Jl.*, **16**, No. 1, 93 (1960)
9. WOLFSON, H. and ELLIOTT, G., U.S. Pat. 2,774,747
10. MATZ, K. R., U.S. Pat. 2,849,631
11. ANON., *Electronics*, **37**, No. 1, 84 (1964)
12. GUL', V. E., SHENFIL', L. Z. and MEL'NIKOVA, G. K., *Soviet Plastics*, **4**, 51 (1966)
13. WHITTIER, F. and LAWN, R. J., U.S. Pat. 2,765,288
14. HARVEY, M. T. and ROSAMILLIA, P. L., U.S. Pat. 2,839,487
15. HARVEY, M. T. and ROSAMILLIA, P. L., U.S. Pat. 2,839,488

16. HOPPER, F. C. and NAPPS, M., U.S. Pat. 2,915,490
17. ARONE, N. F., U.S. Pat. 2,717,216
18. SAFFORD, M. M., U.S. Pat. 2,843,557
19. HONN, F. J., U.S. Pat. 2,904,528
20. HELMREICH, R. F. and HARRY, L. D., *S.P.E. Jl.*, **17,** No. 6, 583 (1961)
21. Dow Chemical Co., Technical Literature
22. CRANKER, K. R. and BRESLAU, A. J., *Ind. Engng. Chem.*, **48,** No. 1, 98 (1956)

CHAPTER 7

Cycloaliphatic and Epoxidised Olefin Resins

7.1 CYCLOALIPHATIC RESINS

The glycidyl ether resins, in the multitude of combinations with curing agents, flexibilisers, fillers, etc., which have been developed, meet most of the performance needs placed on them. However, the more rigorous demands from increasingly complex technologies have led to other types of epoxide resins being synthesised. In particular, the search for new resins has been to obtain:

(a) Systems retaining mechanical and electrical properties on long exposure to high temperatures.

(b) Systems possessing improved electrical properties such as arc and tracking resistance.

(c) Non-yellowing surface-coating resins which retain the high level of performance of the glycidyl ether resins.

Work has tended to concentrate on replacing the aromatic ring of the glycidyl ether resins and on the synthesis of compact molecules which increase cross-link density and impart rigidity to the thermoset resin through cycloaliphatic ring structures. Some examples of the molecular species already examined and finding some industrial use are as follows:[1]

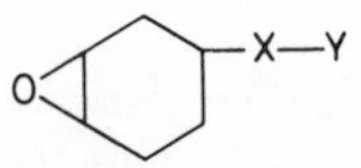

Cyclohexene oxide type

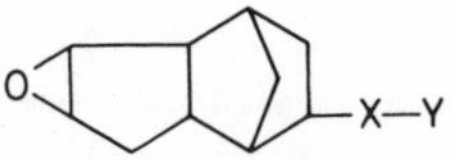

Tricyclodecene oxide type

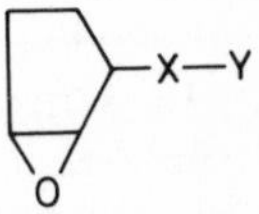

Cyclopentene oxide type

(Y=Hydrogen or the cyclic group already attached to X; X=Ester, ether, acetal, imide, or amide)

The first cycloaliphatics were offered commercially in the U.S.A. in the late 1950s by Union Carbide Corporation, and were 3,4-epoxy-6-methylcyclohexylmethyl-3,4-epoxy-6-methylcyclohexane-carboxylate (structure (a), Fig. 7.1), vinylcyclohex-3-ene dioxide (b),

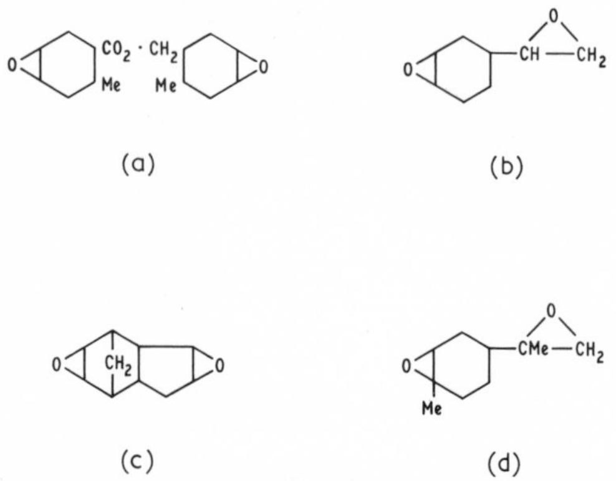

(a)   (b)

(c)   (d)

Fig. 7.1

dicyclopentadiene dioxide (c) and dipentene dioxide (d). Many other groups of polyepoxide compounds have now been synthesised in the search for new resins.

7.1.1 SYNTHESIS OF CYCLOALIPHATIC RESINS

A comprehensive review of this subject by Batzer[1] divides the synthesis into two stages: (*a*) preparation of initial materials, and (*b*) their epoxidation. In stage (*a*) the objective is to obtain compounds having at least one unsaturated ring and a reactive group, or more than one unsaturated group per molecule. Some important synthetic routes to compounds of this type are shown in Figs. 7.2 and 7.3. In Fig. 7.2 the first stage in all cases is a Diels-Alder reaction; the most common dienophiles are acrolein and crotonaldehyde, the diene usually being butadiene. This leads to formation of an unsaturated aldehyde, the common starting point for all subsequent syntheses.

In route (1), two molecules of aldehyde undergo the Tishchenko reaction, in the presence of aluminium isopropoxide. This yields the di-unsaturated ester,[2] which can be epoxidised to form the diepoxide. Route (2) involves reduction of the aldehyde group to methylol, followed by reaction of two molecules of alcohol with one molecule of dicarboxylic acid, yielding the substituted diester.[3] Alternatively, the aldehyde group is oxidised to carboxyl (route (3)) and two molecules of the acid allowed to react with a diol, producing again a substituted diester.[4] The formation of acetals between

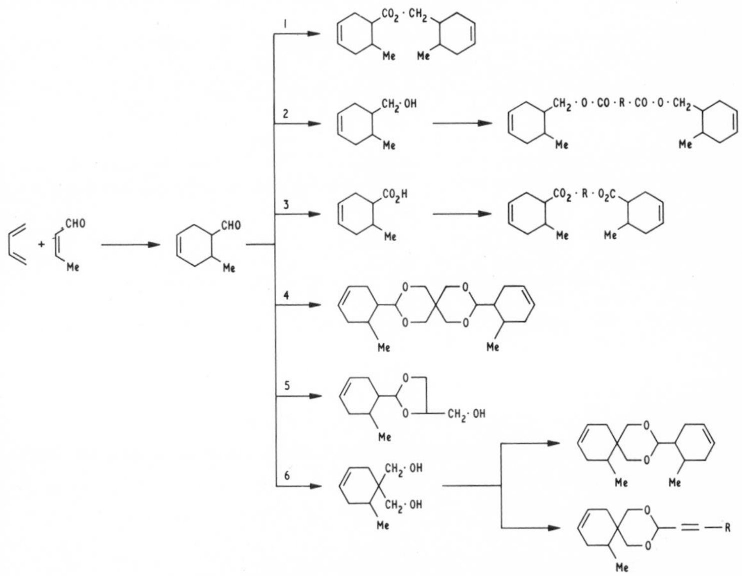

Fig. 7.2 Reaction scheme proposed by Batzer[1]

aldehyde and alcohol is used in routes (4), (5) and (6),[5-9] pentaerythritol is used as the polyol, reacting with two molecules of aldehyde to form the spiro compound in route (4). In route (5) the polyol used is glycerol.

Other synthetic routes are shown in Fig. 7.3, again starting with the Diels-Alder reaction, and using butadiene together with various dieneophiles. Cyclopentadiene and dicyclopentadiene are also readily available starting materials for the synthesis of epoxide resins, the latter being readily epoxidised to form a resin without any further modification.

Gosteva *et al*,[10,11] following the claim of Trigaux,[12] studied the production of epoxide resins for high temperature uses, and reported that those based on epoxidised cyclic hydrocarbons (in particular dicyclopentadiene) had the highest heat resistance. Two of Gosteva's preparations are shown in Fig. 7.4. Libina *et al.*[13] also described the

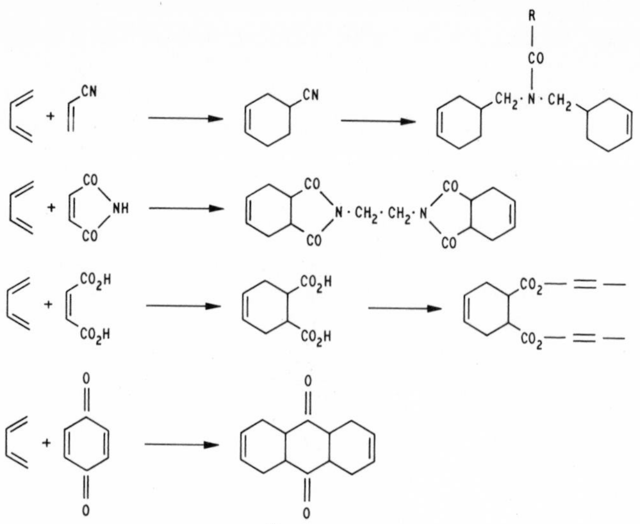

Fig. 7.3

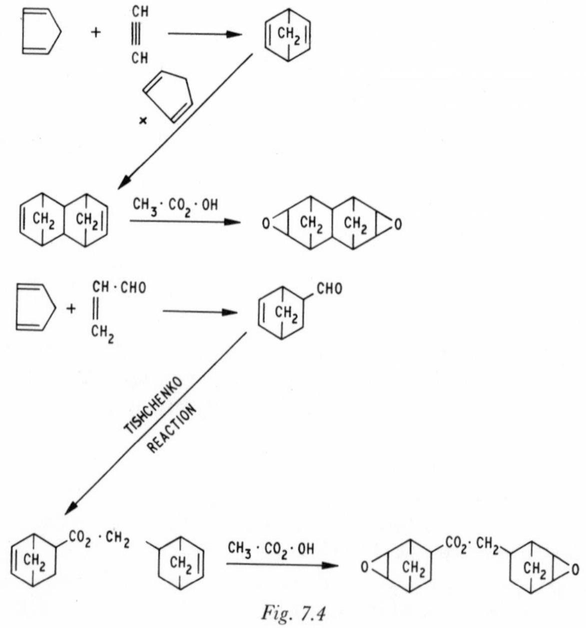

Fig. 7.4

preparation of resins from dicyclopentadiene and its ethers (Fig. 7.5).

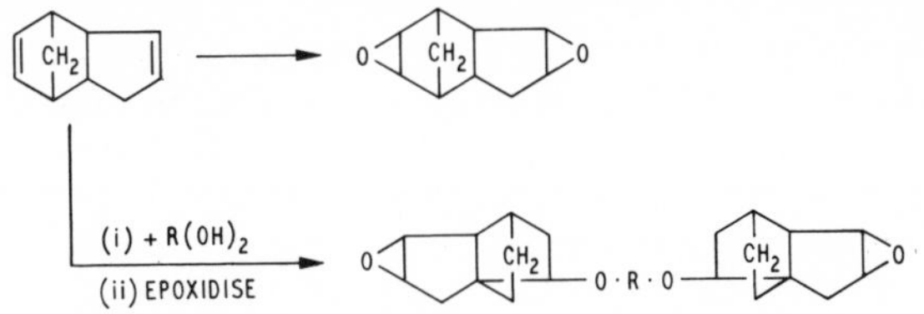

Fig. 7.5

Resins derived from the reaction between bis(2,3-epoxycyclopentyl) ether and hydroxy compounds, catalysed by tertiary amines

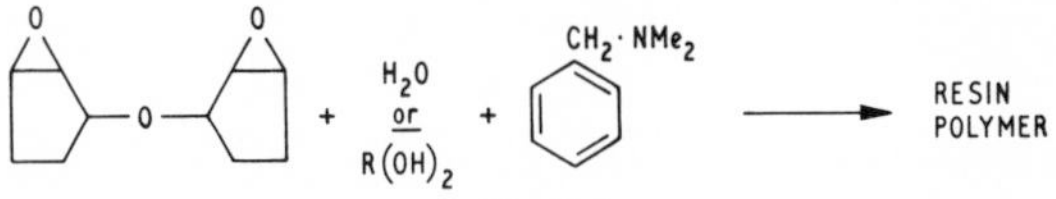

Fig. 7.6

(Fig. 7.6), have been described by Soldatos and Burhans.[14] It is thought that the polymerisation follows an ionic mechanism similar to that of the epoxide-hydroxyl reaction. When the hydroxyl groups are provided by water, a homopolymer is produced, whereas when they are provided by ethylene glycol a copolymer is obtained. These polymers still contain unreacted epoxide groups, which can then be used to produce a thermosetting system, using curing agent MPD.

Burhans *et al.*[15] have also examined 1,4-cyclohexadiene dioxide and 2,3-epoxycyclopentenyl-substituted phenols for their suitability for epoxide resin production.

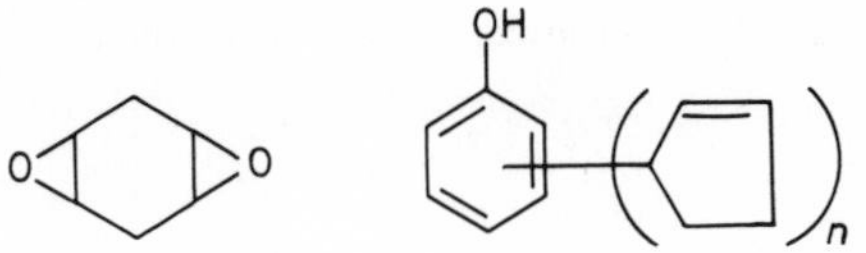

In the second stage of the synthesis of cycloaliphatic resins, the aliphatic double bond must be epoxidised, and all known methods have been summarised by Batzer.[1] However, peracids (in particular peracetic acid) are used in most epoxidations. Peracetic acid is cheaper than the other epoxidising systems and does not give rise to any by-products difficult to remove.

The rate of epoxidation with peracids depends upon the structure of the unsaturated compound and of the peracid itself, together with the reaction temperature and the solvent used. Electron-donating groups in the unsaturated compound tend to accelerate the epoxidation, which enables many epoxidations to be carried out rapidly at temperatures that cause little decomposition of the per-acid. Electron-attracting groups have the opposite effect, slowing down the rate of epoxidation.

Mild epoxidation procedures have also been developed to over-come the undesirable acid-catalysed opening of the epoxide ring, which leads to formation of polyols or hydroxy esters. However, the presence of a small amount of hydroxyl in the resin can be beneficial; as in the curing of the glycidyl ethers, hydroxyl groups act as accelerators.

7.1.2 PROPERTIES OF THE RESINS AND THEIR CURED SYSTEMS

The physical properties of the cycloaliphatics differ from those of the glycidyl ethers, particularly in respect of lower viscosity and lower WPE. Some resins, such as vinylcyclohex-3-ene dioxide, are of sufficiently low viscosity to be used as diluents for the glycidyl ether resins without degrading properties. Some data for resins (a), (b)

Table 7.1 PHYSICAL PROPERTIES OF CYCLOALIPHATIC RESINS[16]

	Resin as in Fig. 7.1 (a)	Resin as in Fig. 7.1 (b)	Resin as in Fig. 7.1 (c)	Liquid diglycidyl ether of DPP
Appearance	Pale straw-coloured liquid	Pale straw-coloured liquid	White powder	Straw-coloured liquid
Viscosity (25°C; cp)	1 200	77	—	10 500
Specific gravity	1·21	1·099	1·330	1·16
Epoxide equivalent	145	76	82	185
Melting point, °C	—	—	184	—
Hardening times at 100°C with:				
(i) Aliphatic polyamine	24 hr	20 min	—	7 min
(ii) HPA + 0·5% BDMA	1¼ hr	45 min	—	45 min
(iii) HPA	15 hr	6¾ hr	—	7 hr

and (c) (Fig. 7.1) are given in Table 7.1. As expected, the cyclo-aliphatics and the glycidyl ethers react at different rates with the various types of curing agents. Lewis[16] demonstrates this by com-paring the hardening times of two cycloaliphatics and a liquid diglycidyl ether with an aliphatic polyamine and an acid anhydride curing agent. Table 7.1 shows the results of this comparison, from which it can be seen that cycloaliphatics tend to react more slowly with aliphatic polyamines than do the glycidyl ether resins. More-over, there is a wide difference in reactivity between the two cyclo-aliphatics themselves. When an acid anhydride is used, particularly with an amine catalyst, there is little or no difference in reactivity between all three resins considered.

Table 7.2 compares the properties of three cycloaliphatic resins with those of a solid glycidyl ether casting resin. The results tend to support the deductions that can be made from an examination of

Table 7.2 PROPERTIES OF CURED CYCLOALIPHATIC RESINS[17]

	Resin as in Fig. 7.7(a)	Resin as in Fig. 7.7(b)	Resin as in Fig. 7.7(c)	Solid glycidyl ether casting resin
Curing agent used	HPA + accelerator			PA
Cure schedule	Gel + 3 hr at 120°C			—
Heat deflection temperature, °C	154	86	150	110
Ultimate tensile strength, lb/in^2	9 600	9 470	9 900	12 000
Ultimate flexural strength, lb/in^2	12 000	13 000	12 900	19 000
Volume resistivity (ohm cm; 25°C)	$>10^{16}$	1.7×10^{15}	$>10^{16}$	6.5×10^{15}
Dielectric constant (60 Hz; 25°C)	3.39	3.40	3.34	—

the molecular structure of the resins. The cycloaliphatics examined are fairly compact ring structures and would be expected to form a rigid and rather brittle structure of high HDT. In addition, the absence of aromatic rings would imply a better resistance to ultra-violet radiation than the aromatic-based glycidyl ethers. Data summarised by Batzer[1] show cycloaliphatic systems to have the following characteristics when compared with the glycidyl ether resins:

(a) Greater compressive and tensile strengths.
(b) Lower flexural and impact strengths.
(c) Lower shrinkage.
(d) Higher HDT and better long-term high-temperature ageing characteristics.

(e) Better retention of electrical properties at high temperatures.
(f) Better arc resistance.
(g) Better resistance to ultra-violet light.

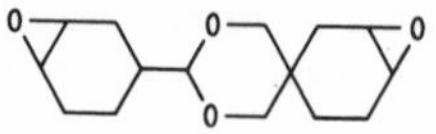

(a)

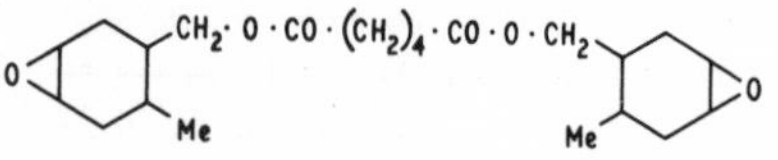

(b)

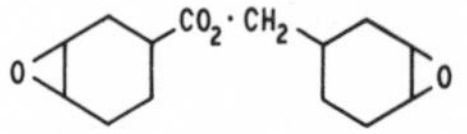

(c)

Fig. 7.7

Dorman[17] has also reported on the properties of cycloaliphatic resins (Fig. 7.7 (a)–(c)), especially with respect to their application in switchgear.

These resins, when hardened with curing agent HPA plus an accelerator, had the properties shown in Table 7.2, but comparison with Batzer's data and with the performance of the glycidyl ether resins is difficult, because of the limited data presented. Dorman also claims that the cycloaliphatic resins, as compared with the glycidyl ethers, have:

(a) Higher temperature performance (up to 250 °C).
(b) Outstanding arc, corona, and tracking resistance.
(c) Low and 'flat' dissipation factor *versus* temperature.
(d) Good outdoor performance under an applied voltage.
(e) Low viscosity, long pot life, and mild temperature curing systems.

Subsequent work by Burhans *et al.*[15] has identified other general types of epoxide resins having strength properties considerably greater than the glycidyl ether types. These new resins are:

bis(2,3-epoxycyclopentyl) ethers
1,4-cyclohexadiene dioxide
(2,3-epoxycyclopentyl)-substituted phenyl glycidyl ethers.

When cured with MPD, all produce castings having compressive moduli and yield strengths much greater than those of the glycidyl ethers. Burhans investigated the family of resins represented by the cyclopentyl-substituted phenols (Fig. 7.8). In general, most of the

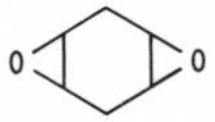

Fig. 7.8

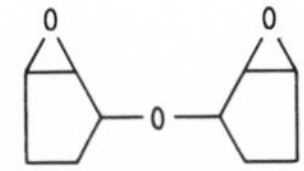

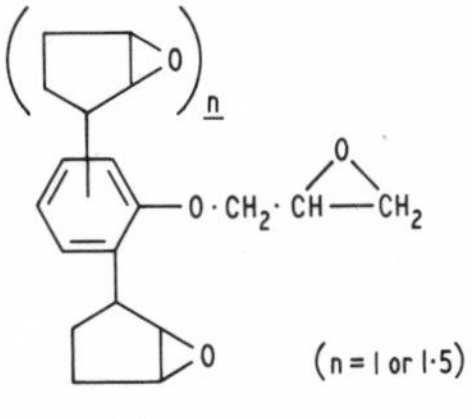

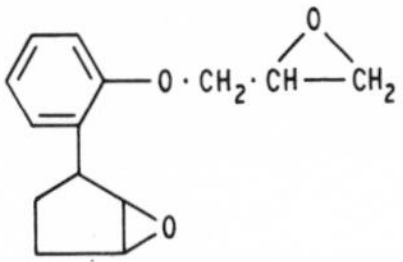

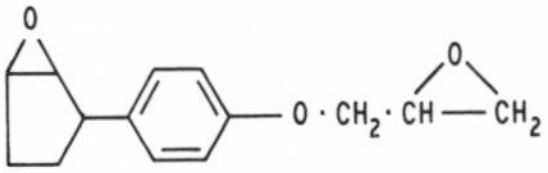

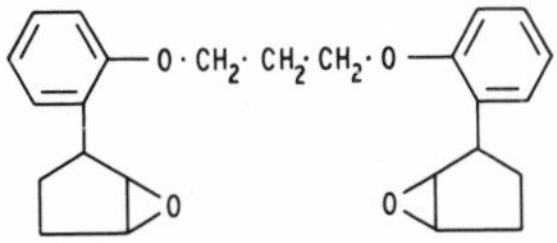

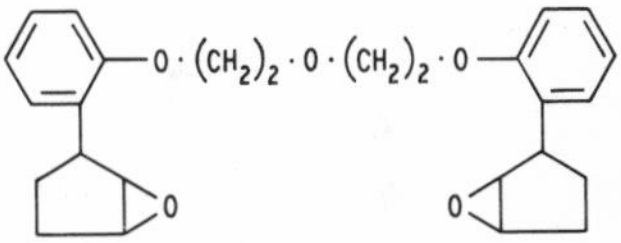

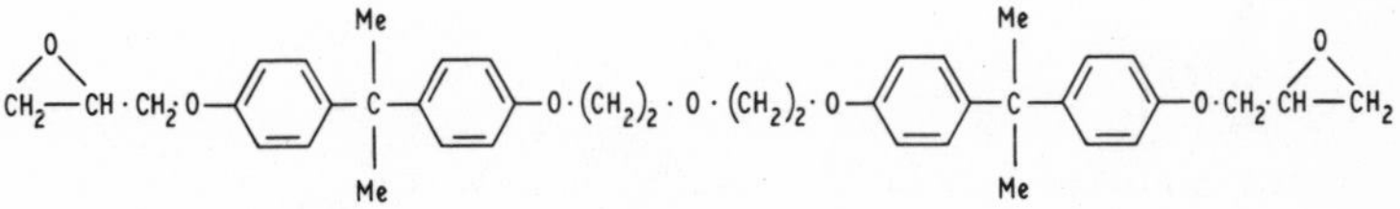

properties of the cast resin systems were a function of the intra-molecular distance between the cross-linking sites on the resin, and the nature of that part of the molecule linking the sites. Figs. 7.9 and 7.10 show graphically the relationship between heat deflection temperature and compressive yield strength, and the distance between the active sites.

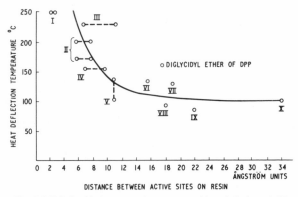

Fig. 7.9 *Relationship between resin structure and heat deflection point*[15]

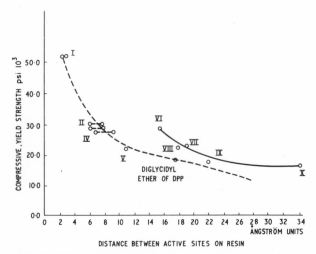

Fig. 7.10 *Relationship between resin structure and compressive yield strength (No data shown for resin III)*[15]

7.2 EPOXIDISED OLEFIN RESINS

These resins are obtained by the epoxidation of polyolefins such as polybutadiene. A schematic representation of one resin of this type, given by Greenspan *et al.*,[18] is shown in Fig. 7.11, and the properties

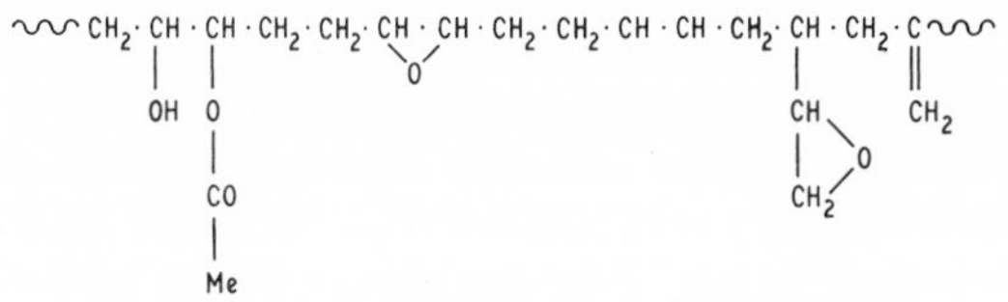

Fig. 7.11

of three resins of this type commercially available in the U.S.A. are given in Table 7.3.

These resins differ from the glycidyl ethers in that they contain more than two epoxide groups per molecule, and in addition have other reactive groups such as hydroxyl and residual double bonds, which can undergo cross-linking and other reactions. Their reactivity

Table 7.3 SOME TYPICAL PROPERTIES OF EPOXIDISED POLYOLEFIN RESINS[19]

	Oxiron * 2 000	*Oxiron* 2 001	*Oxiron* 2 002
Viscosity (poise; 25 °C)	1 800	460	15
Specific gravity	1·010	1·014	0·985
Epoxide, %	9·0	11·0	6·0
Epoxide equivalent	177	145	232
Hydroxyl, %	2·5	2·0	1·9
Iodine number	185	154	—
Range of properties obtained using various curing agents and curing schedules			
Flexural strength, lb/in²	10 000–16 200	15 400–18 100	16 000
Tensile strength, lb/in²	4 000–6 800	5 600–9 700	7 400
Tensile modulus, lb/in² ($\times 10^5$)	1–7·5	—	8·2
Volume resistivity (ohm cm; 25 °C)	1–5×10^{15}	4–6×10^{14}	4×10^{14}
Dielectric constant (60 Hz; 25 °C)	3·17–3·37	3·06–3·34	3·18
Loss factor (60 Hz; 25 °C)	0·006–0·009	0·004–0·016	0·005

*Food Machinery Corporation, U.S.A.

towards the common curing agents is also different. The aromatic-based glycidyl ethers are susceptible to attack by nucleophilic reagents but are not very reactive towards electrophiles, this probably being due to the strong electron-attracting (inductive) effect caused by the ether link attached to the aromatic ring. The epoxidised olefins present the reverse situation; their epoxide groups are more readily attacked by electrophilic rather than by nucleophilic reagents, the initial attack occuring at the oxygen of the epoxide ring. This is probably due to the greater electron density in the carbon-oxygen bond of the ring than in the surrounding carbon-carbon bonds. The reactivity of the linear aliphatic resins is therefore similar to the cycloaliphatics.

The residual unsaturated centres in the epoxidised polyolefins also allow polymerisation with a wide range of monomers such as styrene, and hence permit many different modifications to be made to the resins.

Polyamine curing agents such as TET and MPD are used with an accelerator such as phenol (2 phr) and a high-temperature cure is required. Acid anhydrides, with or without polyols, react rapidly at 90–115 °C, and a variety of properties can be obtained in the cured casting (Table 7.3). In particular, the strength retention at elevated temperatures is good, and this, coupled in one instance with a low viscosity, suggests a possible use in the preparation of laminates.

REFERENCES

1. BATZER, H., *Chemy. Ind.*, No. 5, 179 (1964)
2. FROSTICK, F. C. and PHILLIPS, B., U.S. Pat. 2,716,123
3. PHILLIPS, B. and STARCHER, P. S., U.S. Pat. 2,750,395
4. PHILLIPS, B. and STARCHER, P. S., U.S. Pat. 2,745,847
5. FISCHER, R. F., U.S. Pat. 2,895,962
6. STICKLE Jnr., R., French Pat. 1,234,354
7. BATZER, H. and NICKLES, R., French Pat. 1,233,231
8. BATZER, H., ERNST, O. and FATZER, W., U.S. Pat. 3,072,679
9. PORRET, D., FISCH, W., BATZER, H. and ERNST, O., U.S. Pat. 3,072,678
10. GOSTEVA, O. K., LIBINA, S. L., PRYANISHNIKOVA, M. A., AKUTIN, M. S. and PLATE, A. F., *Soviet Plast.*, **2,** 49 (1962)
11. GOSTEVA, O. K., LIBINA, S. L. and RIVKINA, E. G., *Soviet Plast.*, **3,** 58 (1966)
12. TRIGAUX, G. A., *Mod. Plast.*, **38,** No. 1, 147 (1960)
13. LIBINA, S. L., *et al.*, *Soviet Plast.*, **12,** 18 (1962)
14. SOLDATOS, A. C. and BURHANS, A. S., *Ind. Engng. Chem.*, **6,** No. 4, 205 (1967)
15. BURHANS, A. S., PITT, C. F., SELLARS, R. F. and SMITH Jnr., S. G., 21st Ann. Meeting of Reinforced Plastics Division of SPI, Chicago (1965)
16. LEWIS, R. N., *Trans. Plast. Inst. Lond.*, **31,** 94, 97 (1963)

17.   DORMAN, E. N., Proceedings of 6th Electrical Insulation Conf., U.S.A. (1966)
18.   GREENSPAN, F. P., JOHNSTON, C. and REICH, M., *Mod. Plast.*, **37** 142 (1959)
19.   BRENNER, W., LUMM, D. and RILEY, M. W., *High Temperature Plastics*, Reinhold, New York, 51 (1962)

CHAPTER 8

Surface Coatings

8.1 INTRODUCTION

Surface coatings fall into two groups, (*a*) decorative, and (*b*) protective. The latter category can be subdivided into atmospheric temperature-curing and heat-curing (stoving) systems. A paint consists of a polymeric binder plus pigments, extenders, solvent mixture, and other materials where necessary; a varnish is usually a clear coating formed only from the binder resin. Paints and varnishes containing epoxide resins as binders are used primarily for surface protection, and both room-temperature-curing and stoving systems are available.

The paint must be brought to and spread out on the surface to be painted, and then the binder resin polymerised or allowed to harden to form the film. To transport the paint in this way it must be in a suitable physical form, such as a·free-flowing suspension of solid pigment and filler in a liquid, or as a finely divided powder. In epoxide resin paints the resin and curing agent or other copolymerising resin are usually dissolved separately or together in a strong solvent mixture. In high-solids or solventless systems, the epoxide resin binder is itself a liquid, and little or no solvent is needed to form the liquid paint vehicle. Aqueous emulsions of epoxide resin coatings can also be prepared, but are rarely used in practice. Finally, all of the paint constituents may be solids and can be thoroughly mixed together and made into a finely divided powder.

Methods of applying the paint to the surface are:
(*a*) Brushing
(*b*) Spraying, including hot-spraying

(c) Dipping, roller-coating, flow-coating
(d) Electrophoretic deposition
(e) Fluidised bed and electrostatic or flock-gun spraying for powders.

Once deposited on the surface, the epoxide resin binder is polymerised to a tough film and Fig. 8.1 summarises the chief methods used for this conversion. They are as follows:

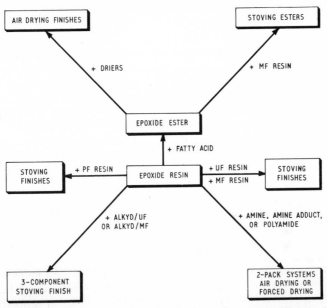

Fig. 8.1 Chief methods of converting epoxide resins into surface coatings. PF = phenol-formaldehyde resin; UF = urea-formaldehyde resin; MF = melamine-formaldehyde resin

1. The resin is cross-linked by polyamines, modified polyamines, or polyamides at room or slightly elevated temperatures.
2. The resin is esterified through its epoxide and hydroxyl groups with suitable oil fatty acids. Depending upon the nature of the fatty acid, the resulting ester can either be polymerised by atmospheric oxygen at room temperature (an air-drying coating) or by heating with or without a copolymerising resin.
3. The resin is blended with another resin such as a phenol-formaldehyde resin, with which it copolymerises on heating (a stoving system).

These basic systems were developed historically for specific reasons, but all form coatings characterised by their excellent adhesion, chemical resistance, hardness, flexibility, and toughness. The amine-cured system was the first to give stoved paint properties from a room-temperature curing paint. Combinations of epoxide with other resins served to upgrade the performance of existing stoving systems, and esters provided a way of obtaining the performance of epoxide resin paints at a lower price. Numerous reviews of the surface-coating applications of epoxide resins have been made, including those by Wheeler,[1] Nickles,[2] Maslow,[3] and Gaynes.[4]

8.1.1 SOLVENTS FOR EPOXIDE RESINS

The solvent mixture is an important part of a paint formulation. Like other synthetic resins, epoxides have a limited tolerance for certain solvents; for example, with a high-M diglycidyl ether, a 40% solution in acetone can be prepared but not a 20% solution. Here the acetone is not regarded as a 'true' solvent, which is defined as one which gives a resin solution which may subsequently be diluted to infinity by the same solvent without resin precipitation.[5]

True solvents for epoxides include methyl ethyl ketone, diacetone alcohol, methylcyclohexanone, and most glycol ethers and their acetates. In cases of limited solubility, the tolerance may be extended by the addition of aromatic hydrocarbons such as toluene, and alcohols such as n-butanol. Although these are not by themselves true solvents for epoxides, mixtures of them frequently are, and are used with the lower-M resins, improving flow and film properties and general solvent balance and lowering the raw material cost of the solvents.

The choice of solvents for a particular system will, to some extent, also depend upon (*a*) the method of paint application, (*b*) the nature of any copolymerising resin, (*c*) any possible reaction between the solvent and the paint constituents, and (*d*) the required viscosity and solids content of the paint.

Epoxide resin esters are soluble in aromatic or aliphatic hydrocarbons depending upon the 'oil length' of the ester, i.e., the relative amount of fatty acid in the ester. Long oil esters are soluble in white spirit and as the fatty acid content decreases, increasing amounts of aromatic hydrocarbons such as xylene have to be incorporated.

8.1.2 COMPATIBILITY

The main film-forming epoxides are the solid diglycidyl ethers of DPP with WPEs of 450–500, 850–1 000, 1 650–2 000 and 2 400–4 000. As M increases, compatibility with other film-formers decreases. Phenol-formaldehyde (PF) resins are often used with the higher molecular weight epoxides in a proportion of up to 40% by weight of the total resin solids, and a wide range of PFs is compatible with epoxides. The urea-formaldehyde (UF) resins are usually compatible with epoxides in all proportions, but the melamine-formaldehyde (MF) resins are less compatible. Both UF and MF resins are used in amounts up to 30% of the toal resin solids.

Many other resins are used with epoxides for surface coatings binders. The more important are alkyds and thermosetting acrylics, which are generally used with the low-M epoxides and with which they have limited compatibility.

8.1.3 PIGMENTATION

Epoxide resin paints may be pigmented with most types of pigment, using normal methods of incorporation. Care must be taken with amine-cured systems in choosing pigments that do not react with the free amine.

8.2 AMINE-CURED SYSTEMS[6–8]

Polymerisation of the resins with polyamines has already been discussed in Chapter 3, and for film formation the polyalkylene polyamines DTA and EDA, polyamides, and amine adducts are the most commonly used. All react readily at room temperature with the liquid diglycidyl ethers and the low molecular weight solid grade of WPE 450–500. Resin and curing agent are dissolved separately in a suitable solvent mixture of ketones, alcohols, aromatic hydrocarbons, and glycol ethers, and if a polyamide is not being used a small amount of a flow-control agent (e.g., a butylated MF resin) is added to eliminate 'pinholing' or 'crawling' of the film.

Cure starts when the resin and curing agent solutions are mixed and the pot life of the paint is usually 1–2 days depending upon the system and the ambient temperature. Fig. 8.2 shows the increase in viscosity with temperature of a solution of a solid resin cured with EDA.

Occasionally, when amine-cured paints are used under cold and humid conditions, the paint film develops a surface bloom or blush. This cloudiness is thought to be due to the free amine in the film

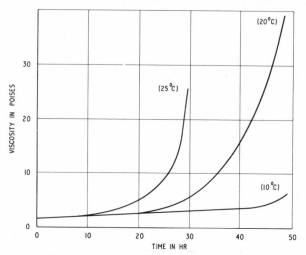

Fig. 8.2 Viscosity increase of an EDA-cured varnish at various temperatures[5]

reacting with carbon dioxide in the atmosphere to form the amine carbonate, which is whitish. Incompatibility between the free amine and the partially cured resin is also a possible source of surface blush. Use of an amine adduct or allowing the mixed paint components to stand for an hour before application overcomes this film defect.

Both the isolated and the *in situ* adducts described in Chapter 4 are used in paint formulations and have more convenient mixing ratios and a lower tendency to surface bloom than the unmodified polyamines. The polyamides derived from dimerised and trimerised vegetable oil fatty acids and DTA are widely used as curing agents for epoxide resin surface coatings.[9-11] These paints are similar in properties to those based on amine adducts and amines and also possess the advantages of a decreased tendency to bloom in the film and non-critical mixing ratios. The resin-to-polyamide ratio can be varied from 75:25 to 50:50 parts by weight. In addition, polyamide-cured films show better water resistance and flexibility and are slower drying.

In general, all of these paint systems develop full properties after 7 days at room temperature, although free epoxide groups remain in the film,[5] and unless a hot cure is used free amino groups will also be present. The minimum practical curing temperature in the absence of any cure-accelerator is 10°C. Amine-cured coatings

Table 8.1 PRINCIPAL DIFFERENCES BETWEEN THE AMINE CURING AGENTS IN SURFACE COATING SYSTEMS[5]

Property	Free amine	In situ adduct	Isolated adduct	Polyamide resin
Blooming	May bloom	Very seldom blooms	Very seldom blooms	Seldom blooms
Ageing	Desirable	Unnecessary	Unnecessary	Unnecessary
Chemical resistance	Excellent	Excellent	Excellent	Improved water resistance, poorer solvent resistance
Flexibility	Very good	Very good	Very good	Excellent
Approx. usage phr*	6	30–35	35–50	33–100
Rate of drying	Fast	Fast	Fast	Slightly slower
Pot life	1–2 days	1–2 days	1–2 days	2–4 days
Nature of curing agent	Volatile and unpleasant	Solution of very low odour	Odourless solid	Semi-liquid of low odour

*Parts by weight of curing agent per hundred parts by weight of resin

combine the advantages of hot-cured finishes with the convenience of air-dried. They have good impact strength and flexibility plus excellent chemical resistance. Table 8.1 summarises the principal differences between the three types of amine curing agents.

8.2.1 COAL TAR MODIFICATION[12]

All amine-cured systems can be modified by the incorporation of compatible coal tar pitches, up to 50–75% of the total binder content. The coal tar acts as a diluent, and when thixotropic fillers are also added very thick high-solids coatings are obtained, with dry film thicknesses of 4–10 mil in one application. Very small amounts of solvents are used and the coatings have similar properties to the unmodified ones, with improved water resistance but reduced acid and solvent resistance.

8.2.2. HIGH-SOLIDS AND SOLVENTLESS COATINGS[13-15]

Paints using only a very small amount of solvent (often called high-solids coatings) or using no solvent at all (solventless coatings) are widely used for structural protection. They enable a thick film to be applied in one application, which reduces time and labour costs, lowers the fire hazard due to solvents, and makes solvent-removal equipment unnecessary.

Certain of these systems use ketimines as curing agents (Section 4.3.4) with a diluted liquid glycidyl ether resin. Any pigments used are mixed with the resin together with a very small amount (1–2%) of a solvent such as isopropanol. On mixing the resin with a low viscosity ketimine, a pot life of about 8 hr at room temperature is obtained, and the paint can be applied by spray or brush. Full chemical resistance properties are developed in a week, and cure requires a damp atmosphere ($>40\%$ RH). The moisture in the pigments also activates the ketimine and speeds cure. Single-coat applications of this system have[5] a direct impact strength of 48 in/lb and Bucholz hardness of 45. On immersion for 4 months in a range of substances such as distilled water, 5% sulphuric acid, 15% hydrochloric acid, 20% sodium hydroxide, aviation gasoline, ethanol, and toluene, no failure of the film was observed.

8.2.3 APPLICATIONS OF THE AMINE-CURED SYSTEMS

The ability of amine-cured epoxide resin coatings to provide outstanding chemical resistance and physical properties, without the need for stoving, has led to their very wide use as protective coatings for many different substrates.[16] The protection of structural steelwork and of processing equipment, chemical plant, oil refineries, and food factories are typical examples of the use of these coatings in corrosive environments. Metal or concrete marine, road, rail or static storage tanks are frequently coated internally to prevent corrosion and to preserve the purity of the contents. The coatings will withstand wine, beer, cider, molasses, alkalis, detergents, and crude and refined oil products. Similarly, drums used for oil products and chemicals are often coated internally.

The excellent alkali resistance of these coatings makes for an outstanding paint for concrete, cement, asbestos-cement, and plaster surfaces. Concrete walls can be decorated and then given severe scrubbings with detergents for cleaning purposes without any breakdown of the film over many years, or concrete buildings can be provided with a weather-resistant finish.[17]

The amine-cured system is also employed as a protective coating for paper book covers and fatty food wrappings and as a finish for bar tops, tables, floors, and other wooden surfaces.

An important recent development has been their use as the resin binder in zinc-rich coatings. These coatings, which usually consist of 92–95% zinc dust and 8–5% binder, have been used for the protection of steel for many years. The binder has usually been a thermoplastic material and the coating has frequently required. overcoating to improve properties such as water resistance. However, loss of inter-coat adhesion was often experienced, accompanied by a softening of the primer. The use of an epoxide resin binder has overcome these difficulties and much improved the corrosion resistance properties of the system. In addition, these primers can be applied to steel plate or to structural steel work immediately after blast cleaning, preventing any further corrosion from taking place on storage. This primer does not limit the way in which the metal can be handled subsequently; it can be cut and welded in the usual way, and is readily overcoated with a wide range of paints. The value of epoxide-resin-based zinc-rich primers has been acknowledged by the shipbuilding industry, particularly in the Netherlands, Japan and the U.K. Corrosion is a major problem in the marine field and Klaren and Russell[18] report an estimate of the cost of unchecked corrosion in an oil tanker fleet of 200 vessels amounting to £2·8 m. per annum. They also give a vivid example of the effect of corrosion on the hull of a tanker causing a significant increase in power needed to maintain the same ship speed.

Any of the adduct- or polyamide-cured types of coating can be used in conjunction with the low molecular weight solid resin, and Vogelzang[19] describes a primer which employs the very high molecular weight phenoxy-type resin as a binder for 85:15 flake zinc: flake aluminium, at 75:25 pigment-to-binder ratio which has improved through hardening and flexibility.

In the marine field, coal-tar-modified amine-cured systems are widely used for hull paints, tank lining, and superstructure paints.[20]

8.3 EPOXIDE RESIN ESTERS

The solid diglycidyl ethers of DPP possess secondary hydroxyl groups at intervals along the polymer chain as well as terminal epoxide groups. They are therefore able to react with long chain drying (i.e., possessing conjugated unsaturation) or semi- and non-drying fatty acids, forming esters, the epoxide group also esterifying under these conditions, being equivalent to two hydroxyl groups. In principle, resin grades not possessing hydroxyl groups can be esterified, although this is rarely done in practice. Clearly, a wide range of epoxide resin esters is possible, the variables being the resin grade, the type of acid used, and the extent to which the resin is esterified. Valuable general accounts of these esters have been given by North[21] and Wheeler.[22,23]

The process of esterification consists of heating the resin and fatty acid together to about 240 °C with or without a solvent and maintaining the mixture at this temperature until the esterification is complete. This is judged be measuring the acid value of the mixture, which is then cooled and a solvent such as xylene added, to produce a solution of 50–60% wt. solids content. The important properties of an ester are its acid value, which generally should be as low as possible, and its viscosity. High acid value can result in inferior drying properties and poor paint performance. Reproducible low viscosity for an ester is extremely important and can be a critical factor in a process such as roller coating. Paints are usually formulated to a given viscosity, and if one of the components of the paint varies in viscosity, then at constant paint viscosity the solids content must vary from batch to batch. This could have serious effect on the film thickness and spreading power of the paint.

Reactions which can occur in the esterification process are illustrated diagrammatically in Fig. 8.3. Reaction (a) is the acid-epoxide reaction, forming an ester plus a new hydroxyl group. This is a faster reaction than (b), which is the usual hydroxyl-carboxyl reaction to form an ester plus water. Reaction (c), the epoxide-hydroxyl reaction, causes resin homopolymerisation, one epoxide group being consumed, but the overall hydroxyl group concentration remaining unchanged. Reaction (c) also results in a considerable increase in the viscosity without any lowering of the acid value of the system. This branching reaction has already been discussed in Chapter 2, when it was mentioned that it tends to occur when the resin is heated above 200 °C in the presence of a base. Reaction (d) is the homopolymerisation of the fatty acid if it is of the drying type, via the double bonds in the long chain.

The esterification reactions are all catalysed by bases, which convert the carboxyl group into a carboxylate ion. These anions react more rapidly with hydroxyl and epoxide groups than do the unionised acid and other epoxide groups. It is therefore essential in the preparation of epoxide resin esters to ensure that sufficient basic catalyst is present. Reed[24] examined eight additives as esterification catalysts and found that small amounts of certain additives, especially sodium benzoate and zirconyl compounds, influenced to a

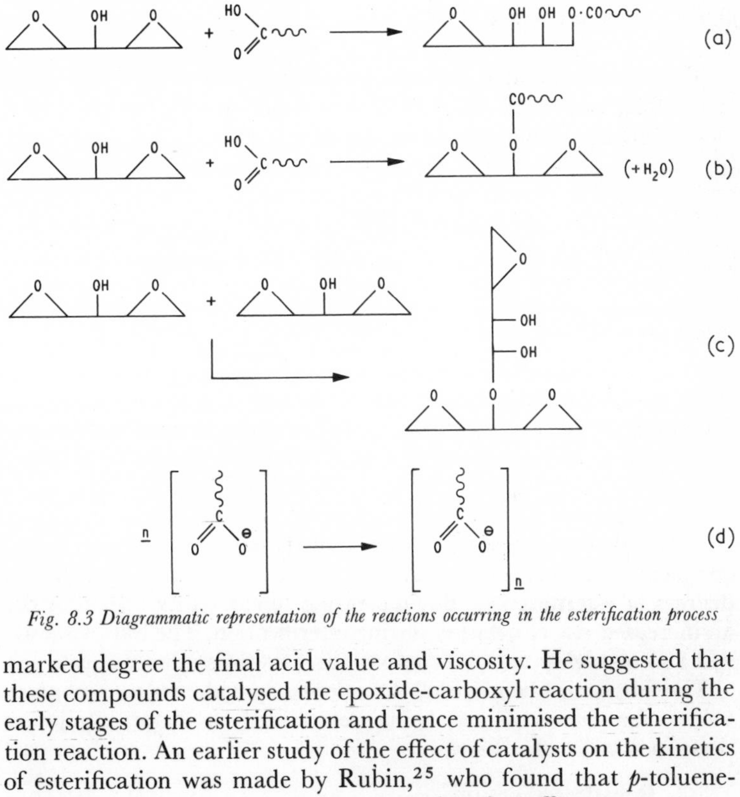

Fig. 8.3 *Diagrammatic representation of the reactions occurring in the esterification process*

marked degree the final acid value and viscosity. He suggested that these compounds catalysed the epoxide-carboxyl reaction during the early stages of the esterification and hence minimised the etherification reaction. An earlier study of the effect of catalysts on the kinetics of esterification was made by Rubin,[25] who found that *p*-toluene-sulphonic acid gave the maximum accelerative effect.

If insufficient base is present, reaction (c) (Fig. 8.3) leads to a large increase in viscosity, consuming epoxide groups and slowing down the consumption of acid which has fewer of these groups with which to react, and must therefore react in the slower esterification process (b). Heating times are therefore prolonged and this aggravates the situation, since reaction (d) then also commences, leading to viscosity increase. The net result is a high viscosity ester with an unacceptably high acid value.

Resin base content is not the only factor that influences ester viscosity. The structure and molecular weight distribution of the resin also have an effect on resin performance in the esterification process and can also influence ultimate ester performance.

8.3.1 RESINS AND ACIDS USED

The resin most used in ester preparation has $M c.$ 1 400, WPE 870–1 025, and m.p. 95–105°C; but other grades are also used, depending upon the end-use for the ester. ·

Resin	Approx. value of n in general formula	Main use in esterification
m.p. 64–76°C WPE 450–500	2·0	Very long oil esters, i.e., 90% esterification or above, and where low viscosity required
m.p. 95–105°C WPE 870–1 025		All esters up to 80–90% esterification
m.p. 125–132°C WPE 1 650–2 025		Short and medium oil length esters (up to 50% esterification of semi- and non-drying oil acids)

The higher the molecular weight of the resin, the greater is the chemical resistance and toughness of the film, assuming constant degrees of esterification. But increasing functionality also provides an increased risk of gelation during esterification. The ester viscosity will also be higher and the compatibility towards other resins lower.

The important acids used to esterify epoxide resins are given below, the choice of acid being determined by factors similar to those in alkyd technology.

Dehydrated castor Linseed Soya Safflower Tall oil Rosin	Drying and semi-drying acids
Coconut Lauric Castor	Non-drying acids

The term 'oil length' in epoxide resin technology is taken to be the degree of esterification of the epoxide resin with the fatty acid, short indicating 30–50%, medium 50–70%, and long oil length 70–90% esterification. The long oil esters are used for air-drying brushing paints, whereas the medium and short oil esters are used for air-drying spraying systems, especially for stoving primers and finishes in conjunction with UF and MF resins.

8.3.2 ESTER MANUFACTURE

Two methods are used:

(a) *The fusion process*. Resin and acid are heated together either in an open pot or in a closed kettle. The open pot is somewhat primitive and is not recommended, since it yields esters of high viscosity and bad colour. The closed kettle allows an inert gas atmosphere to be maintained in the kettle, which excludes oxygen and helps water removal, leading to paler coloured esters.

(b) *The azeotropic process*. To the resin and acid is added about 2·5% wt. of xylene to help remove the water by a xylene-water azeotrope, and an inert gas blanket is employed together with efficient stirring in a closed kettle. This method is the easiest to control and provides the palest coloured esters. The equipment needed is, however, more complex, a condenser and water separation trap being required to remove the water from the azeotrope that distils over and to allow the xylene to be returned to the kettle. In addition a powerful stirrer is fitted. In operation the fatty acid is usually placed in the kettle first and preheated to 150 °C; the resin is then added slowly with stirring, together with the xylene. The mixture is heated to full cooking temperature (240–260°C) as quickly as possible, care being taken to control any excessive frothing that may occur when the azeotrope is boiling off by means of careful stirring and control of the heating system.

The reaction is followed by means of periodic determinations of the acid value (as mg KOH per g sample) and viscosity of the mixture, and the temperature maintained at the top cooking temperature (i.e., highest temperature reached) until the appropriate values for these parameters are reached. The reaction mixture is then cooled and sufficient solvent such as xylene or a low-aromatic white spirit (depending on the oil length of the ester) is added to obtain a solution containing about 50% solids.

The cooking temperature used will depend upon the acid employed and the oil length, acid value, and viscosity desired. For a given acid value the lower the cooking temperature the lower the viscosity. This is illustrated by Figs. 8.4 and 8.5, which show the

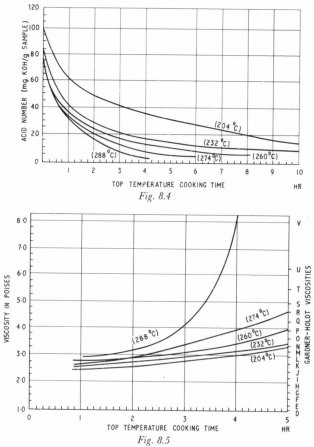

Fig. 8.4

Fig. 8.5

effect of cooking temperature on acid value and viscosity for a soya fatty acid ester. In general, acid values are less than 10.

The final ester solution is filtered to remove haziness and is then ready for use. Air-drying esters and those being stoved without any amino resin require the addition of metallic driers, usually cobalt naphthanates to provide 0·04% cobalt metal on ester solids, but with stoving esters these are not necessary when UF and MF resins are the co-reactants.

Two typical epoxide resin esters are as follows:

	Long oil linseed ester	Dehydrated castor oil short oil ester
Epoxide resin (WPE 870–1 025)	43·5% wt.	60% wt.
Linseed oil fatty acids	56·5% wt.	—
Dehydrated castor oil fatty acids	—	40·0% wt.
Cooking temperature	240–260°C	240°C
Approx. cooking time	5–6 hr	3–5 hr
Acid value (mg KOH per g sample)	7–10	1–2
Solvent	White spirit	Xylene
Viscosity (50% solids; poises at 25°C)	5–8	3·5–4·5

An alternative method of preparing epoxide resin esters, developed in the U.S.A. by Union Carbide Corporation,[26] starts with a liquid resin, DPP, and the fatty acid, and can produce a very wide range of esters. Two reaction steps are involved:

(a) In the first step the liquid resin and DPP react together in the presence of a lithium catalyst such as lithium naphthenate, to form a polymer of higher M. The terminal epoxide groups of this polymer are then esterified with the calculated amount of fatty acid, forming a terminally esterified resin plus two further hydroxyl groups.

(b) This intermediate ester is then esterified further by allowing the secondary hydroxyl groups along the molecular chain to react with a further quantity of fatty acid.

8.3.3 STYRENE-MODIFIED AND VINYLTOLUENE-MODIFIED ESTERS

Epoxide resin esters may be modified by the slow addition of styrene or a vinyltoluene, plus a polymerisation catalyst such as di-t-butyl peroxide, during the cooking process.

Styrenation reduces the cost of the ester and in addition improves drying speed, colour, and colour retention, but reduces solvent resistance. The vinyltoluene modification offers, in addition, solubility in aliphatic hydrocarbon solvents, which allows the ester to be used in brushing quick-drying maintenance paints and roller-coating applications. The styrenated esters find use in industrial stoving primers and finishes.[27,28]

8.3.4 USES FOR EPOXIDE RESIN ESTERS[22, 23]

Esters are often employed in applications where alkyd or tung oil-phenolic systems are not suitable, or where a better performance from the paint film is necessary. Epoxide resin ester finishes have better adhesion, chemical resistance, flexibility, and hardness, as compared with the conventional coatings.

Air drying esters. Long oil linseed esters are widely used as brushing air-drying maintenance primers, especially where mildly corrosive conditions are encountered, as in chemical plant, oil refineries, breweries, and steel plants. They do not possess the outstanding chemical resistance of amine-cured systems, but are superior to alkyd finishes in this respect, providing a paint film with a much increased life. One characteristic of ester coatings is their tendency to lose gloss on exterior exposure more quickly than alkyd finishes. This loss of gloss is due to the slow 'chalking' of the film, i.e. loss of pigment from the surface of the coating. Whilst this makes the esters unsuitable for gloss-retaining top coats it helps to minimise dirt retention on the coating and hence assists the latter to remain clean. Chalking does not have any deleterious effect on the performance of the paint, and after five years of exterior exposure it has been difficult to measure the actual resulting decrease in film thickness. Air-drying esters have also been used on railway carriages and for floor and spar varnishes over timber. Short oil air-drying esters such as the dehydrated castor oil esters have been used for aircraft finishes and external drum paints.

Stoved esters. Short or medium oil length drying oil esters can be stoved by themselves to form films possessing good flexibility, adhesion, and chemical resistance. They are widely used for metal primers and for collapsible tube external coatings such as toothpaste tubes, where their flexibility allows the tube to be crushed without damaging the film. The short oil length dehydrated castor oil ester is the one most commonly used, small amounts of driers being added, such as 0·005% wt. of cobalt metal on resin solids, with a stoving schedule of 20–30 min at 120–150 °C. The use of styrenated and vinyltoluene-modified esters as industrial primers and finishes has already been mentioned.

Significant improvements in hardness, flexibility, and chemical resistance of stoved ester films can be made by incorporating a proportion (usually not less than 30% wt.) of a melamine-formaldehyde (MF) resin. These ester/MF paints are used as domestic appliance primers, usually as the lauric ester, or as drum coatings.

For the brightest whites and best colour retention non-drying esters are preferred, such as lauric esters. For maximum hardness and chemical resistance, however, a drying oil ester, usually the dehydrated castor oil, is widely used. Driers are not normally used in the ester/MF systems and stoving schedules are in the range 20–40 min at 120–150 °C.

Probably the largest use for stoving esters is as automobile primer/surfacers. Car painting throughout Europe follows the same basic pattern. After degreasing and phosphatising of the metal, a primer is applied either by spray, dip, or slipperdip. This primer can contain an epoxide resin as a three-component system with an alkyd and a UF or MF resin; it is followed by one or more coats of a primer/surfacer, which in most cases is an epoxide resin ester. After filling and sanding, either two or three finishing coats are applied. These are often of the alkyd-MF type, but there are many variations. Thermosetting acrylic systems are becoming very popular, and thermoplastic acrylics are also used.

The difficulties of poor penetration of paint into inaccessible areas, together with the difficulty in coating edges of panels and the runs and sags that occur when solvent-borne or water-dispersible primers are applied by dipping, have led to the development of a new method of painting, namely electrophoretic deposition.[29–31] This method is similar to electroplating in that the object being coated is dipped into a bath and becomes an electrode as in a cell. The electrolyte is an aqueous solution of a soluble resin system and a second electrode is placed in the bath. On passing a DC current, the paint system (i.e., pigment plus soluble binder) is deposited on the object by a process of electrophoresis, electrolysis, and finally electro-osmosis. The main advantages of this method of paint application have been summarised by Walker[32] as follows:

(a) Good control of film thickness,
(b) deposition of paint on the whole surface of the processed article including edges,
(c) freedom from runs and sags,
(d) efficient utilisation of paint,
(e) no solvent washing in box sections,
(f) no flash-off time required, and
(g) no expensive solvents required and fire hazards reduced.

Epoxide resin esters have now been developed which are water thinnable and can be applied by electrodeposition. Van Westrenen et al.[33, 34] have described the two main methods of preparing the water-soluble esters, both of which achieve water-thinnability via carboxylate ion formation.

One technique involves the maleinisation (i.e., the chemical addition of maleic anhydride through its unsaturated group, to unsaturated fatty acid chains) of a long oil epoxide ester. The resin, a low molecular weight solid glycidyl ether, WPE 450–525, is esterified with excess linseed oil fatty acids by the azeotrope method until 85–90% conversion of hydroxyl groups is attained (AV 30–34). The ester is then heated with maleic anhydride to 240 °C for 1–3 hr, and ethylene glycol monobutyl ether added to the mixture after cooling, followed by neutralisation with 25% ammonia. The resulting ester is water soluble. The high conversion of hydroxyl to ester groups is necessary to avoid the anhydride reacting with these resin hydroxyls.

The second method consists of reacting short oil esters with phthalic anhydride through their residual hydroxyl groups, forming half-esters which are also polycarboxylic acids. On neutralising with ammonia, a water-soluble ester is obtained.

8.4 STOVING SYSTEMS

Epoxide resins, and especially the solid diglycidyl ether having a WPE of 1 650–2 000, can be used to improve the performance of other synthetic film-formers such as PF, MF, and UF resins. The PF resins are mostly low-M butylated resoles containing phenolic hydroxyl plus etherified and unetherified methylol groups. At stoving temperatures of 180–200 °C the phenolic hydroxyls react with epoxide groups, and both types of methylol react with the secondary hydroxyls of the epoxide resin.[35] The net result is a cross-linked polymer network which possesses the best chemical resistance and physical properties of all epoxide coating systems.

To prepare a coating of this type it is usually sufficient to blend together solutions of the two resins, with pigments if required. The paint can them be applied by brush or spray, and the coated article heated for 20–30 min at 200 °C to copolymerise the resins. Cure temperature can be reduced to 175 °C by using an acidic accelerator such as salicylic or p-toluenesulphonic acid.

Urea-formaldehyde (UF) resins copolymerise with epoxide resins at 180–200 °C, and a wide range of butylated UF resins is compatible with epoxides. The epoxide:UF ratio is usually about 70:30 by weight, and clear or pigmented systems are made by simple cold-blending of the resin solutions. The stoving schedule required to achieve a fully hardened film is similar to that for the PF system, namely 20 min at 180–200 °C, which can be lowered to 30 min at only 150 °C by using a small amount of acidic accelerator.

A limited number of MF resins is compatible with epoxides, and coatings based on them are prepared and used in a similar way to the PF and UF systems.

Epoxide-MF systems have good colour, hardness, and heat resistance, and are used for industrial stoving paints, especially where light colours are needed, as in domestic appliances. The epoxide-UF system is also used similarly as a stoving primer or top coat for industrial or domestic equipment,[36] and for can and drum coatings and wire enamels.

The extremely high chemical resistance shown by the epoxide-PF system has led to its use as an internal protective coating for metal food containers, storage tanks, and pipelines, and in general as a protective coating against corrosive products such as detergents, solvents, chemicals, and fuels.[37]

8.4.1 COMBINATIONS WITH ALKYD OR THERMOSETTING ACRYLIC RESINS

Conventional alkyd-MF stoving enamels can have their properties improved by replacing half of the alkyd by an epoxide resin. This three-component system has better adhesion, flexibility, and chemical and solvent resistance than the unmodified enamel and is therefore used for white and pastel finishes for domestic appliances. The alkyd used is often a short oil glycerol, or one based on tri-methylolpropane, the epoxide usually having WPE 450–500, and m.p. 64–76 °C.

The solid diglycidyl ethers of DPP are also polyols, and can be used to replace part of the glycerol or other polyol used in alkyd resin manufacture, with consequent improvement of hardness and chemical resistance.

Thermosetting acrylic resins may possess free amino and carboxylic acid groups from monomeric acrylamide and acrylic acid used in their preparation. These functional groups will react with the epoxide group at 150 °C and above, in the presence of a basic catalyst,[38,39] and hence the two resins may be copolymerised. This combination of epoxide with acrylic resin had much improved properties as compared with the unmodified acrylics. Adhesion, gloss, flexibility, and other mechanical properties are all enhanced, together with the resistance of the paint film to a wide range of corrosive materials such as alkalis, detergents, salt water, and distilled water.

Because of compatibility difficulties the epoxide resins used are limited to liquid and low-M solid grades, whilst most thermosetting acrylics are suitable. The epoxide resin is used at about 10–25% by weight of the total resin solids; cure schedules vary, but a typical recommendation is 30 min at 170°C. These epoxide-acrylic resin coatings are widely used as one-coat systems for refrigerator and washing machine top-coats.[40]

8.5 EPOXIDE RESINS CURED WITH ISOCYANATES

The isocyanate group (—NCO) reacts very readily with groups such as carboxyl and hydroxyl, containing active hydrogen atoms, a reaction which is the basis for the formation of polyurethane polymers. These are widely used for flexible and rigid foam manufacture[41] and also for paints and lacquers.

The secondary hydroxyl groups present along the molecular chain of the solid epoxide resins can also react readily with the isocyanate group, thus:

$$R \cdot NCO + HO— \quad \rightarrow \quad R \cdot NH \cdot CO \cdot O—$$

Hence, the reaction between a polyisocyanate and an epoxide resin will lead to rapid cross-linking.[42] The polymer formed is useful as a surface coating, possessing very high chemical resistance. A two-can system is therefore possible; the resin, which may contain between 8 and 12 hydroxyl per molecule, is dissolved in a suitable solvent mixture and kept separate from the isocyanate solution until the paint is to be used. This is similar to the amine-cured paints discussed in Section 8.2. All the constituents of this coating system must be dry, and the solvents must not contain hydroxyl groups, to avoid reaction with the isocyanate.

Coatings based on the epoxide-isocyanate system are usually hard-dry two hours after application at room temperature, and the advantage of the system is that it possesses the same outstanding physical properties as the amine-cured systems in addition to superior acid-resistance.

The epoxide groups of the resin may not enter into the reaction with the isocyanate,[43] but can be converted into hydroxyl groups which certainly will react. This modification is usually carried out by using a dialkanolamine or a polyethylene glycol. In the case of the alkanolamine, the amino hydrogen opens the epoxide group forming an hydroxyl group and attaching to the terminal carbon atom of the resin two additional hydroxyl groups linked through a nitrogen atom, thus:

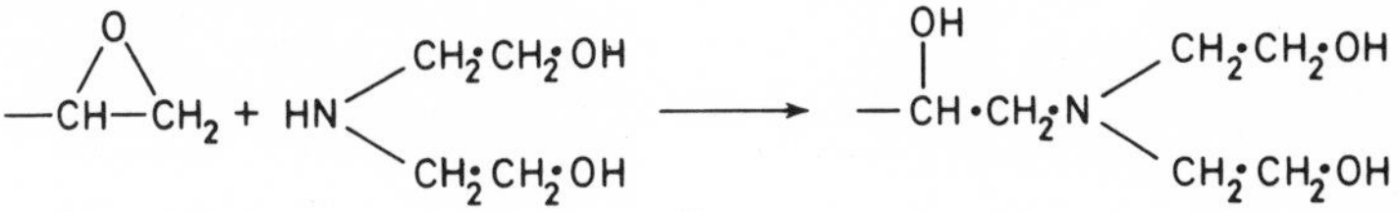

8.6 EPOXIDE RESIN POWDER COATINGS

Thermoplastic powder coatings based on PVC, nylon, polythene, and some less important polymers, have been known for many years, but the development and use of thermosetting powders based on epoxide resins is a more recent phenomenon. These powders are 100% solids, contain no liquids, and are dry dust-like mixtures. They are applied by electrostatic spraying or fluidised bed, and very thick films can be applied at one application. Absence of solvents also reduces the risk of fire, always present with conventional paints and lacquers. Useful reviews of these powders have been made by Sprackling[44] and Stephens.[45]

An epoxide powder usually contains, in addition to the resin, a curing agent, flow-control agent, pigments, and fillers. The choice of each item depends upon the method of manufacture of the powder and also the manner of application. Formulations using a liquid resin, blends of liquid and solid resins, and single or blended solid resins have all been used in conjunction with curing agents such as DDM, DICY, accelerated amines, and BF_3 complexes.[5,46] The dry components are blended together using one of the many techniques described in the literature, such as mixing in a twin-screw extruder, in a Z-blade mixer, or in a ball mill. In the first two cases the cooled mass obtained after mixing is broken up and reduced by grinding, first to 6–10 mm mesh size, and then to the required particle-size distribution by further grinding. Dry blending in a ball mill achieves mixing and particle-size reduction in one operation, although the materials should have a particle size of 6 mm maximum when being charged into the mill.

Blending in a Z-blade mixer involves heating the materials to 90–100 °C for up to $\frac{1}{2}$ hr, and it is therefore necessary to use latent curing agents which do not react at this temperature. A typical preparation of a DICY (dicyandiamide) powder has been described[5] as follows:

The resins (Epikote 1001 75 parts by wt., Epikote 1004 25 parts by wt.) are premelted together at 130–150 °C in a Z-blade mixer and 3–5 parts by wt. of a flow control agent melted in. Pigments and fillers are then added (0–60 parts by wt.) and the temperature allowed to fall so that the viscosity increases and allows dispersion to occur at maximum shear. Thixotropic filler (2·5 parts by wt.) and DICY (4·4 parts by wt.) are finally added and dispersed by running the mixer for a further 15 min at 90–100 °C.

The melt is discharged into trays, the mixer temperature being taken up to 130 °C if necessary. When cold the mass is broken up and powdered in a suitable high-speed mill of the pin disc type. The powder is then sieved and any powder that is retained on a 45 mesh ASTM sieve (350 μ) is returned for further grinding.

For this type of coating a cure schedule of 30 min at 200 °C is recommended, although powder coatings which cure rapidly at lower temperatures are available and widely used.

There are four main methods of applying powders, as follows:

(a) *Electrostatic spraying.* The powder, finely ground, is given an electrostatic charge as it passes through the spray gun in an air stream. The article to be coated is earthed and the charged powder particles are attracted to it, forming an even coating over the surface. Thin films are obtained when the article is coated cold. If it is preheated, thick films, at least 20 mm (500 μ), can be obtained. After spraying, the coated article is heated in an oven to cure the powder. Bright[47] has given a full description of the electrostatic spray gun.

(b) *Fluidised bed method.* The powder is fluidised in conventional equipment and the article to be coated is heated and immersed in the powder. The powder particles, on coming into contact with the hot object, fuse and adhere to it. The coating is then stoved in an oven.

(c) *Electrostatic fluidised bed method.* The fluidised powder is given a charge by means of electrodes placed in the bed. The earthed article to be coated is then passed over the surface of the bed and attracts the charged particles on to itself. The coating is then stoved.

(d) *Flock spraying.* The article is preheated and the powder sprayed on to it through a conventional wide-orifice gun. The partially fused coating is finally cured in an oven. The powders can also be sprayed through a flame gun, where the powder is melted as it passes through the flame. On striking the cold surface the molten particles adhere and coalesce to form a continuous coating.

Epoxide powder coatings can be formulated so as to have excellent chemical resistance, impact resistance, and electrical properties, these being chiefly determined by the nature of the curing agent. Tables 8.2, 8.3, and 8.4 give some chemical, mechanical, and electrical properties of three different powder formulations. Because of this high level of properties, plus the ease of applying thick single

Table 8.2 CHEMICAL RESISTANCE PROPERTIES OF SOME EPOXIDE RESIN POWDER COATINGS[5]
(*thickness* 5×10^{-3} in; 125 μ)

Reagent	*Powder based on DICY. Stoved 30 min at 200°C*	*Powder based on BF$_3$ complex. Stoved 30 min at 180°C*	*Powder based on DDM. Stoved 30 min at 150°C*
18 months immersion at room temperature			
10% Acetic acid	Fair	Good	Good
20% Nitric Acid	Good	Good	Excellent
20% Ammonium hydroxide	Good	Poor	Good
Ethanol	Excellent	Good	Excellent
200 hr immersion at 90°C			
Distilled water	Good	Fair	Good
5% Sulphuric acid	Poor	Good	Fair
10% Sodium hydroxide	Fair	Good	Good

Table 8.3 MECHANICAL PROPERTIES OF SOME EPOXIDE RESIN POWDER COATINGS[5]
(5×10^{-3} in; 125 μ *coatings on mild steel*)

Property		*Powder based on DICY. Stoved 30 min at 200°C*	*Powder based on BF$_3$ complex. Stoved 30 min at 180°C*	*Powder based on DDM. Stoved 30 min at 150°C*
Adhesion		Excellent	Excellent	Excellent
Impact resistance	(direct)	Excellent	Excellent	Excellent
	(reverse)	Excellent	Fair	Good
Flexibility		Excellent	Poor	Excellent

Table 8.4 ELECTRICAL PROPERTIES OF SOME EPOXIDE RESIN POWDER COATINGS[5]

	Cure schedule	Film thickness 5×10^{-3} in	Breakdown voltage (RMS)
Powder based on DICY	30 min at 200 °C	16	8·8 kV
Powder based on BF$_3$	30 min at 150 °C	22	16·7 kV

(Breakdown voltages measured under oil with $1\frac{1}{2}$ in dia. electrode and 30 kV per min voltage rise)

	Temperature °C	Dielectric constant 1 kHz	Power factor 1 kHz	Volume resistivity (ohm cm)
Powder based on BF$_3$.	20	4·2	0·007	5×10^{15}
Cured 1 hr at 150 °C	50	4·4	0·009	4×10^{14}
	75	4·6	0·026	2×10^{12}
	100	6·0	0·078	9×10^{8}
Powder cured	20	4·1	0·007	1×10^{16}
3 hr at 150 °C	50	4·1	0·007	5×10^{14}
	75	4·2	0·015	3×10^{13}
	100	5·0	0·094	2×10^{10}

(Values determined on $\frac{1}{8}$ in-thick sheets)

coats to complicated and irregularly shaped objects, epoxide powders can be used in many widely differing applications. Examples are coatings for electric motor stators and armatures, capacitors and resistors, pipelines (interior and exterior), coils and springs, tubular metal goods, kitchen tools, etc.[48,49,50]

REFERENCES

1. WHEELER, R. N., *Paint Technol.*, **18**, 207, 131 (1954)
2. NICKLES, O. L., *Off. Dig. Fed. Paint Varn. Prod. Clubs*, **27**, 185 (1955)
3. MASLOW, P., *Off. Dig. Fed. Paint Varn. Prod. Clubs*, **30**, 277 (1958)
4. GAYNES, N. I., *Ind. Finish. J.*, **35**, No. 8, 101 (1959)
5. Shell Chemical Co., Technical Literature
6. OTT, G. H. and ZUMSTEIN, H., *J. Oil Colour Chem. Ass.*, **39**, 5, 331 (1956)
7. ZONSVELD, J. J., *J. Oil Colour Chem. Ass.*, **37**, 670 (1954)
8. ZONSVELD, J. J., *Farbe Lacke*, **60**, 431 (1954)
9. KEENAN, H. W., *J. Oil Colour Chem. Ass.*, **39**, 5, 299 (1956)
10. GLASER, D., FLOYD, D. and WITCOFF, H., *Off. Dig. Fed. Paint Varn. Prod. Clubs*, **29**, 385, 159 (1957)
11. WITCOFF, H., *Am. Paint J.*, **42**, No. 6, 90 (1957)
12. HERZBERG, M., *Paint.-Pigm.-Varn.*, **35**, 383 (1959)
13. ZONSVELD, J. J., *Farbe Lacke*, **64**, 125 (1958)

14. ZONSVELD, J. J., Paper to 4th FATIPEC Congress (1957)
15. ZUMSTEIN, H., *Fette, Seifen, Anstrichmittel,* **60,** No. 7, 547 (1958)
16. ADOMENAS, A., *J. Oil Colour Chem. Ass.,* **48,** No. 5, 423 (1965)
17. SOMERVILLE, G. R., *Off. Dig. Fed. Paint Varn. Prod. Clubs,* **37,** 487, 921 (1965)
18. KLAREN, H. and RUSSELL, G. C. C., *Shipp. Wld.,* **154,** No. 3747, 643 (1965)
19. VOGELZANG, E. J. W., *Paint Mf.,* **34,** No. 1, 41 (1964)
20. BANFIELD, T. A., *Off. Dig. Fed. Paint Varn. Prod. Clubs,* **36,** 477, 1133 (1964)
21. NORTH, A. G., *J. Oil Colour Chem. Ass.,* **39,** No. 5, 318 (1956)
22. WHEELER, R. N., *Paint Technol.,* **19,** 212, 159 (1955)
23. WHEELER, R. N., *Paint Technol.,* **19,** 215, 260 (1955)
24. REED, F. E., *Am. chem. Soc. Symp.,* Atlantic City (1959)
25. RUBIN, W., *J. Oil Colour Chem. Ass.,* **35,** 386, 418 (1952)
26. WYNSTRA, J. and KURJY, R. P., Brit. Pat. 873,252
27. HEAVERS, M. J., *Paint Mf.,* **28,** No. 1, 5 (1958)
28. HEAVERS, M. J., *Paint Mf.,* **28,** No. 2, 48 (1958)
29. BURDEN, J. P. and GUY, V. H., *Trans. Inst. Metal Finish.,* **40,** No. 3, 93 (1963)
30. GLOYER, S. W., HART, D. P. and CUTFORTH, R. E., *Off. Dig. Fed. Paint Varn. Prod. Clubs,* **37,** 481, 113 (1965)
31. TASKER, L. and TAYLOR, J. R., *Surface Coatings,* **1,** 10, 378 (1965)
32. WALKER, I., *Surface Coatings,* **1,** No. 10, 386 (1965)
33. VAN WESTRENEN, W. J., WEBER, J. R., SMITH, G. and MAY, C. A., Paper to 8th FATIPEC Congress (1966)
34. VAN WESTRENEN, W. J. and TYSALL, L. A., *J. Oil Colour Chem. Ass.,* **51,** No. 2, 108 (1968)
35. BRUIN, P., *Chemy. Ind., Lond.,* **1,** 616 (1957)
36. GLASER, M. A., BROMSTEAD, E. J. and WEAVER, G. L., *Off. Dig. Fed. Paint Varn. Prod. Clubs,* **27,** 3 (1955)
37. DUNN, P. A., *Paint, Oil Colour J.,* **122,** No. 2820, 988 (1952)
38. PIGGOTT, K. E., *J. Oil Colour Chem. Ass.,* **46,** No. 12, 1009 (1963)
39. ROBINSON, P. V. and WINTER, K., *J. Oil Colour Chem. Ass.,* **50,** No. 1, 25 (1967)
40. STEWART, D., *Paint Technol.,* **30,** 7 19 (1966)
41. PARKER, D. V. B. and PHILLIPS, L. N., *Polyurethanes,* Iliffe, London (1962)
42. DETSCH, H., *Farbe Lacke,* **66,** 74 (1960)
43. POSWICK, J. and DRAMAIS, C., *Off. Dig. Fed. Paint Varn. Prod. Clubs,* **30,** 407, 1431 (1958)
44. SPRACKLING, J. M., *Surface Coatings,* **1,** No. 3, 80 (1965)
45. STEPHENS, T. J., *Surface Coatings,* **1,** No. 3, 90 (1965)
46. HUMPHREYS, K. W., *J. Oil Colour Chem. Ass.,* **47,** 6, 434 (1964)
47. BRIGHT, A. W., *Product Finish.,* **15,** No. 3, 54 (1962)
48. BAILLIE, G. H., *Insulation, Lake Forest,* **12,** No. 10, 58 (1966)
49. HOMER, H. H., *Insulation, Lake Forest,* **12,** No. 9, 30 (1966)
50. MALLIA, J. C. and CROCE, A., *Insulation, Lake Forest,* **13,** No. 2, 42 (1967)

BIBLIOGRAPHY

Industrial Paints; Basic Principles, TYSALL, L. A., Pergamon, Oxford (1964)
Epoxidharzlacke, WEIGEL, K., Wissenschaftliche Verlagsgesellschaft, Stuttgart (1965)

CHAPTER 9

Epoxide Resins in the Electrical Industry

9.1 INTRODUCTION

In the early days of the electrical industry, insulation was achieved by the use of materials such as mica, glass, and porcelain, and also by fibres such as cotton and silk when more flexibility was desired, as in wiring. The properties of rubber were quickly exploited and this became a standard insulation material, both in its flexible form and after vulcanisation to ebonite. Because of the bulk occupied by wires covered by rubber or textile yarn, wire enamels based largely on phenolic resins were developed. Enamel-covered wires enabled more turns to be wound per unit volume on, for example, the stator coil of a dynamo, which obviously increased efficiency because of the higher flux densities obtainable coupled with the higher working temperature.

Equipment such as transformers was and often still is insulated, and at the same time protected from moisture, by coating or embedding the component in asphalt or bitumen. Such procedure, is, however, not very satisfactory mechanically, since the high-softening bitumens are brittle at room temperature and progressively soften and tend to flow as the component becomes warm. In addition, where impregnation of windings has been carried out, electrical efficiency is impaired.

Paraffin waxes have been in use for many years for insulation, particularly in small components such as condensers for radio equipment. Here again these suffer from the limitations of the low softening point of waxes and mechanical weakness.

During and after World War II it became increasingly apparent that waxes and bitumens were no longer suitable for potting or encapsulation in many electrical applications. Aircraft and rocket-propelled missiles now required delicate electrical equipment of increasingly complicated design which at the same time had to withstand previously unheard of mechanical shocks at a variety of temperatures and pressures. They also needed to be resistant to moisture, corrosive chemicals, and (under humid tropical conditions) mould growths.

Phenolic resins and more recently polyester resins have been and are used in some instances to replace bitumen and wax. Phenolic resins, while often used as mouldings and in laminates, are not very suitable for potting or encapsulation because of high shrinkage on cure (8–10% on volume) and a high power factor (often over 0·1) which gives rise to undesirable power losses and consequently to overheating. Polyester resins, whilst having reasonable electrical properties, suffer from high shrinkage on and after cure. High after-cure shrinkage sets up internal stresses which can give rise to cracking and also can affect the electrical characteristics of a circuit, particularly if coils or condensers are involved.

Silicone resins possess very good insulation and dielectric properties which are well maintained at elevated temperatures, but they possess very poor mechanical strengths, need high cure-temperatures for development of optimum properties, do not adhere very well to many substrates, and are fairly expensive.

Epoxide resins, however, provide a class of materials which although not possessed of such good electrical properties as the silicones is nevertheless superior in this respect to phenolics and polyester resins. In addition they are mechanically strong with a low cure-shrinkage, have good adhesion particularly to metals, are resistant to mechanical and thermal shock, and possess higher resistance to moisture, chemical attack, and mould growth.

Some of the earliest uses for epoxide resins were in the electrical and electronics industry, the resins being used for casting the insulation of small transformers and similar equipment and in the encapsulation of electronic components. The use of epoxide casting resins has grown steadily over the years, and developments in formulations now enable them to be used for dipping, sealing, impregnation, transfer moulding, bonding, laminating, and surface coating.[1,2,3] Many of these terms are self-explanatory, but it is of value to note the specific meanings of some.

Casting involves pouring the resin and curing agent mixture into a suitably designed mould, the cured resin system taking on the shape of the mould from which it is removed when cure is complete. In *encapsulation* the resin system is poured into a mould in which is mounted a component such as a transistor assembly and its associated network. After cure, the unit is removed from the mould, and the components are thus obtained enveloped in resin, with any components such as coil windings, or laminations in a transformer, impregnated by the resin. *Potting* is essentially the same as encapsulation, except that the mould forms an integral part of the completed unit, for example in the form of an aluminium case. In the *dipping* process no mould is employed, and the component is simply dipped into a thixotropic resin system and cured, thus obtaining a hard resin coating which can be regarded as an encapsulation process though here little or no impregnation of windings is achieved.

9.2 PROPERTIES AND CHARACTERISTICS OF CASTING SYSTEMS

The epoxide resin system, containing curing agent, fillers, and other necessary ancillary materials, is usually liquid at the processing temperature, and a number of important factors must be considered when deciding whether any specific formulation is to be useful in casting, potting, and encapsulation. These are:

(a) *Viscosity*. The system must be low enough in viscosity to fill voids, and penetrate the electrical components and windings that are being potted or encapsulated. Many electrical components contain paper separators or porous insulation which must also be thoroughly impregnated. In addition, a low viscosity assists the removal of air from the system. Viscosity can be reduced by heating or by the use of a reactive diluent, but since diluents frequently have significant vapour pressures, and cause an overall lowering of properties, the use of heat to reduce viscosity is preferred. This will of course be accompanied by decreasing pot life and increasing reactivity, together with more rapid exothermic heat evolution from the system, and these factors have been discussed in a previous Chapter.

(b) *Absence of volatiles*. Good castings are free from entrapped bubbles, which if present become focal points for crack propagation, and internal discharges. Considerable care is therefore taken to free resins, hardeners, and fillers from volatiles and air which could form voids in the casting. Certain resins and curing agents have high vapour pressures which may cause a significant portion of them to be removed accidentally during vacuum treatment used to remove air bubbles from the mixture.

(*c*) *Exotherm.* The resin system should cure with a sufficiently low exotherm to enable castings of the required size to be prepared and also to ensure that the peak temperatures reached in the casting will not damage electrical equipment.

(*d*) *Shrinkage.* When a resin is cured, shrinkage occurs in the liquid state and also during the change from liquid to solid during gelation. The pre-gelation shrinkage is taken up by headers which top-up the resin mix; the post-gelation shrinkage is the part that introduces stresses which can lead to damage in sensitive inserts, change the electrical characteristics of a component, or lead to the formation of internal stresses with subsequent cracking of the casting. Parry and Mackay[4] measured the pre- and post-gel volume shrinkage of a number of casting systems, by a dilatometer technique, and found that the shrinkage/time curve differed markedly from one type of amine curing agent to another. Different types of amines gave different total volume shrinkages, but showed less pronounced differences for the post-gelation shrinkage. In general the results were highly reproducible, post-gelation volume shrinkages of about 2–3% being observed. Total volume shrinkages (pre- and post-gel phases) were about 4–6%. Ways in which shrinkage can be minimised include the use of flexibilised resin systems, the incorporation of fillers, and careful control of the curing process.

(*e*) *Thermal expansion.* A low coefficient of thermal expansion of the cured resin is desirable to minimise stress due to differential expansion or contraction between the encapsulated component and the surrounding resin during temperature changes. Here again the use of fillers assists in lowering the coefficient of thermal expansion.

(*f*) *Crack resistance.* Perhaps the most important characteristic of an epoxide resin casting system for electrical use is its resistance to cracking on severe thermal cycling. Little difficulty is encountered unless inserts such as conductors or iron cores are included in the casting, in which case the differential contraction is important. The coefficients of expansion of copper and iron are 17×10^{-6} and 12×10^{-6} per degC respectively, whilst that of an unfilled epoxide resin casting is 60×10^{-6} per degC. There is a significant difference in the contraction, and this is the prime source of cracking. This difference can be reduced by the use of fillers, silica flour reducing the coefficient of expansion to about 32×10^{-6} per degC. It is thus the invariable practice to use fillers when making large castings. Flexibilised resin systems are not used to any extent to modify the stresses imposed by the differential contraction, since the ultimate tensile strength is reduced and is reached before the material has elongated sufficiently to take up the stress. The current approach is to design carefully, on a molecular level, the curing agent and resin components in order to obtain a cross-linked polymer network which can combine both high tensile strengths and a degree of flexibility.

There are many ways in which the resistance of epoxide resin systems to thermocycling can be determined, and these usually involve the encapsulation of a metal insert, such as a hexagonal nut or toothed washer. The insert is usually completely embedded just beneath the surface of the casting, in such a way that any cracks that occur can be transmitted to the surface of the casting and observed visually. The casting is subjected to a thermal cycling schedule, which at its extreme would involve a casting being taken from $-75\,^{\circ}C$ to $+170\,^{\circ}C$ extremely rapidly.

(g) *Thermal conductivity*. This is an important factor since the heat produced by electrical components that have been encapsulated must be dissipated through the resin.

(h) *Operating environment.* It is obviously essential for the resin system to have acceptable electrical properties in its final cured form and for the whole system to be able to withstand the operating environment. This usually involves working above room temperature, and also controlled conditions of humidity, chemical attack, and exterior weathering.

9.3 LARGE CASTINGS

The use of epoxide resins for cast insulation was one of the earliest applications of these versatile materials. Utilising their excellent insulation and mechanical properties, coupled with the ease of casting complicated and accurate shapes in a simple operation, epoxides soon became accepted by the electrical industry for use in current and voltage transformers and various types of bushing.

Starting in Switzerland in 1947–1948 with resins produced by CIBA, the technique soon spread to other European countries, including the United Kingdom, and now is well established, consuming a considerable tonnage of resin per annum. The resin used almost exclusively until the last two years has been a solid diglycidyl ether having WPE 330–380 and m.p. 50–55 °C. It is used with 30 phr of phthalic anhydride as hardener and 200 parts by weight of silica flour as filler. The system is chosen for large castings because of its good insulating properties, strength, low exothermic heat, and toughness over a wide temperature range, helping to ensure crack-resistant castings.

The minimum curing conditions are 16–24 hr at 120–130°C, longer times leading to some improvement in properties but also slowing down the 'turn-round' time of the expensive moulds. The maximum cure temperature is dependent on the mass of the casting; 0·5 kg castings can be cured at 150°C, but 500 kg castings at no more than 120°C. Heat evolution and the degree of shrinkage on cooling are the chief limiting factors. Gradual heating up to cure temperature and slow cooling after cure is essential.

Table 9.1 TYPICAL PROPERTIES OF CURED CASTINGS OF THE SOLID CASTING RESIN SYSTEM[5]
(PHTHALIC ANHYDRIDE USED AS CURING AGENT)

	Unfilled	*Filled* 200 phr silica flour
Specific gravity at 20°C ,	1·2–1·3	1·7–1·8
Thermal conductivity, cal/cm sec °C	$4·6–4·8 \times 10^{-4}$	$15·4–15·6 \times 10^{-4}$
Coefficient of thermal expansion, per degC	$60–65 \times 10^{-6}$	$30–35 \times 10^{-6}$
Water absorption (BS 2782; %wt.)	0·085–0·095	0·030–0·035
Heat deflection temperature, °C (ASTM 648; 264 lb/in²)	100–110	105–115
Ultimate tensile strength, lb/in² (ASTM D638)	9–12 000	9–12 000
Compressive yield strength, lb/in² (ASTM D695)	16–18 000	26–32 000
Ultimate flexural strength, lb/in² (ASTM D790)	16–20 000	16–20 000
Modulus of elasticity, lb/in² (ASTM D790)	$4–6 \times 10^5$	$1–2 \times 10^6$

Table 9.1 shows some typical properties of the cured system, with and without filler, whilst Fig. 9.1 (a)–(g) depict the variation in certain electrical properties with frequency and temperature for the same systems.[5]

It is especially important in large castings to minimise the rate of heat evolution. As mentioned previously (Chap. 6) the ratio of surface area to volume is a critical factor, particularly in large castings, and high internal temperatures can cause bubbling and void formation together with cracking of the casting on cooling.

These considerations lead to the virtual exclusion of most poly-

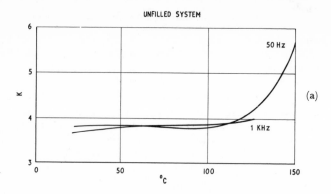

UNFILLED SYSTEM

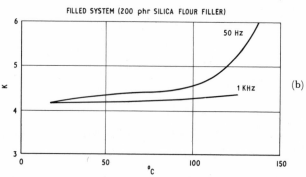

FILLED SYSTEM (200 phr SILICA FLOUR FILLER)

Fig. 9.1 Electrical properties of the solid casting resin/PA system.[5] *(a) and (b) Variation of dielectric constant with temperature*

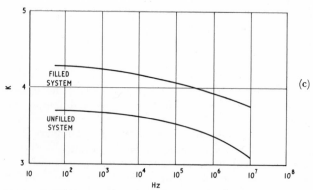

Fig. 9.1 (c) Variation of dielectric constant with frequency

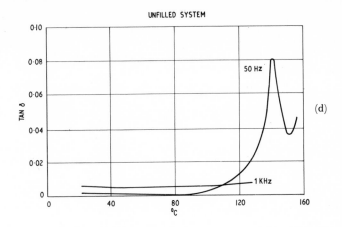

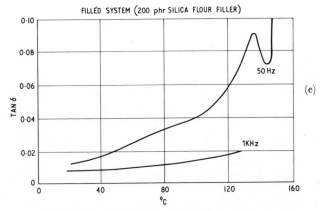

Fig. 9.1 (d) and (e) Variation of power factor with temperature

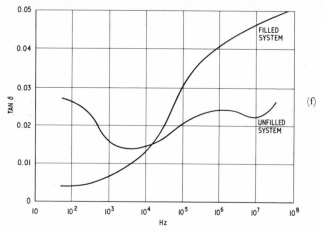

Fig. 9.1 (f) Variation of power factor with frequency

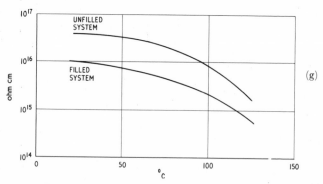

Fig. 9.1 (g) Variation of volume resistivity with temperature

amine curing agents for large casting use because of their high re-activity towards glycidyl ether resins. Acid anhydrides are preferred since they have lower reactivity, and hence lower exotherms are produced. In addition, anhydrides have the advantage of giving low viscosity mixtures with resins and longer pot lives and lower shrinkage than with amine curing agents. When cured, the castings have excellent electrical and mechanical properties, are light coloured, and can be formulated to have high temperature-resistance. A disadvantage of anhydride curing agents is the long and high temperature cure schedules needed, although accelerators assist in overcoming this limitation.

The wide use of phthalic anhydride with the solid casting resin described earlier has been chiefly due to its being the first in the field and a cheap material giving acceptable performance properties. Other anhydrides such as THPA, HPA, and NMA are now becoming used for large castings with the solid resin or with a low-M semi-solid or liquid glycidyl ether resin.

9.3.1 CASTING PROCESS

The casting process has been well described in the literature and consists, at its very simplest, of heating the various ingredients of the system, mixing them together, and pouring them into a mould. This basic technique, although widely used for small castings where the inclusion of air bubbles does not matter, is not suitable for castings where bubbles cannot be tolerated or where complex shapes are being cast. In these cases elaborate vacuum-degassing techniques are used, the most widely employed method being to mix and pour the resin system under vacuum.

Fig. 9.2 is a diagram of a typical casting plant suitable for batch processing. When in operation with the solid resin/PA system a pre-weighed amount of resin and dried silica flour is placed in the mixing chamber. The temperature is then raised to 140–150 °C and the mixture stirred for 2–3 hr under a vacuum of 0·1–0·5 mm mercury. This process ensures complete wetting-out and dispersion of the filler and the removal of entrapped volatiles, moisture, and air. After mixing the temperature is dropped to 125–130 °C and the vacuum removed.

Concurrently the hardener is weighed and melted in a separate chamber at 130 °C and added at the appropriate time to the blend of resin and filler. The hardener is not placed under vacuum, since phthalic anhydride readily sublimes and excessive loss can occur through volatilisation. The mix is stirred under vacuum for a further 10–20 min at 130 °C to remove any air that has been introduced, but at this stage the pot life of the system is about 1–1¼ hr and the second stage of vacuum mixing is therefore relatively short. When it is completed, the casting mix is drawn into moulds which have been previously heated and treated with a release agent, in the evacuated (1–10 mm mercury) casting chamber.

Once the casting has been carried out the vacuum is released and the filled moulds removed to an oven for curing.

In pilot or batch-casting plants, manual working is required in all phases of the operation and mixing chamber capacities vary from 1 kg up to 500 kg. In the larger equipment two mixer units are often

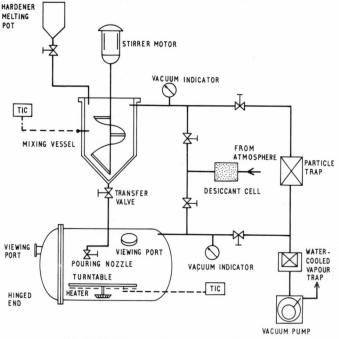

Fig. 9.2 Diagram of resin casting plant[5]

installed for the resin and filler blending in order to speed up production. The greater and more widespread use of epoxide resins has, however, led to the development of continuous casting plants where the metering and mixing of the components can be made fully automatic.

Although the solid casting system is widely used, it has some distinct limitations. These are:

1. The hardener (phthalic anhydride) sublimes and can block flow-lines in the vacuum equipment.

2. Production times could be greatly reduced if the mixing, evacuating, and curing stages were speeded up.

3. The relatively high viscosity of the resin at the casting temperature limits the amount of filler that can be incorporated to 200 parts for every 100 parts of resin. A higher filler loading would lead to distinct cost savings in the system as well as lowering exotherm and producing lower stresses in the casting. It could also lower the thermal expansion coefficient nearer to that of metals.

These disadvantages in the solid system have led to the solid resin and silica flour being sold ready-mixed in a given ratio. This eliminates the lengthy mixing stage at the casting plant, although the mixture needs to be heated under vacuum to remove air and for mixing with the curing agent.

A different approach is to use a lower molecular weight semi-solid diglycidyl ether resin with an anhydride curing agent which does not sublime (THPA). This system is claimed to have considerable handling and cost advantages over the solid resin, together with almost identical electrical and mechanical characteristics and much improved crack-resistant properties. A further development is the marketing of a mixture containing resin, curing agent, and filler all mixed in the correct proportions. It is supplied as polythene-wrapped slabs which are fed on to an inclined-plane melter, the molten mixture running directly into the mixing chamber of the casting plant. The user then need only deaerate the mixture before pouring.

9.3.2 USES FOR LARGE CASTINGS

Epoxide resin castings are widely used in power transmission equipment, especially where high voltages are concerned, for bus bar insulation, bushing insulators, and switchgear insulation. Rothwell[6] has reviewed the use of cast epoxides in high voltage switchgear and given examples of the wide use of epoxide resin insulation, in both air-break and oil-break indoor switchgear, outdoor and flameproof switchgear, and current and voltage transformers. He also comments on the advantages gained from using resin insulation in the many different units described. Certain equipment can be newly designed to a more compact shape, often with a significant reduction in overall dimensions. Dorman[7] also provides a useful review of the use of epoxide resin cast insulation in switchgear and lists 18 typical uses for cast resin insulation, including cable boxes, oil-tight terminal boards, bushings, drive shafts, and 11 kV current transformers. In addition, large sections of electrical machines, generators, and traction motors all contain epoxide resin insulation.

A notable exception, but a very important potential market for the resins, lies in outdoor insulation. At present, at least in Europe, most outdoor insulators (e.g., on power pylons) are made from porcelain or glass. However, porcelain insulators chip easily and are expensive, since in their manufacturing process they need to be machined to an accurate shape. In addition, unlike epoxide resin castings, metal inserts cannot be included at the manufacturing stage and assembling an array of three or more ceramic insulators requires considerable mechanical fixing. Cast resin is also much stronger in tensile, flexural, and impact strengths, has superior dielectric properties, and is one-third of the density of porcelain. Similar considerations apply to glass.

Resin insulators would offer improvements in most respects if they were able to withstand outdoor exposure to ultra-violet light and humidity. The most important technical factor which has prevented their outdoor use is their tendency to surface breakdown by crazing under these conditions. In situations of high surface electrical stress and polluted atmospheres, surface cracks trap dirt particles and allow electrical surface tracking. Once this short-circuit path has been established the resin carbonises along it when a current passes, setting up a permanent tracking path.[8] At low electrical stress and atmospheric pollution, glycidyl ether resin cast insulators have performed well,[9] but under severe conditions their performance has been unacceptable.

Much research has been devoted to (a) identifying special fillers which inhibit tracking, and (b) developing new resin systems with improved weathering characteristics. The most promising filler to emerge from this work is trihydrated alumina, investigated by the General Electric Company, U.S.A. Their work, published in 1963, reports outstanding resistance to tracking failure by a cast insulator in a glycidyl ether resin system filled with trihydrated alumina when used in 69 kV, 115 kV, and 500–525 kV transmission lines.[10]

The new resin systems considered have all been of the cyclo-aliphatic type described in Chapter 7, which when hardened with a curing agent such as HPA give a system with improved tracking properties.[11,12] This is probably due to the absence of aromatic rings from both resin and curing agent, which greatly reduces the tendency to carbon formation when the resin is pyrolysed. When a current passes the likelihood of carbon formation is also decreased, and the possibility of forming a permanent track is reduced.[13] Cycloaliphatic systems also show an improved resistance to ultra-violet light, and this reduces crazing and surface breakdown with consequent reduction in dirt retention on the insulator. This improved weathering characteristic is again no doubt due to the absence of aromatic rings in the resin and curing agent structures. Although results so far have been disappointing, no doubt research will continue in an effort to find a suitable system for outdoor insulators.

9.4 SMALL CASTINGS, POTTING, AND ENCAPSULATIONS

In the production of small castings, pottings, and encapsulations, it is not especially difficult to control the exotherm produced, and shrinkage is also less of a problem than with large castings. The range of resin systems that can be used to produce crack- and void-free casts is therefore wide and will include polyamines as well as anhydride curing agents in conjunction with liquid diglycidyl ether resins.[14]

For small castings, aliphatic polyamines such as TET are often used with a liquid resin, frequently plasticised by the addition of up to 15 phr of dibutyl phthalate. This system will polymerise satisfactorily at room temperature, although properties are improved by a post-cure at elevated temperatures. The system is limited to fairly small articles because of the exotherm developed, and castings usually have a maximum HDT of 55–65 °C or 80–90 °C when the plasticiser is omitted. Formulations with longer pot lives, but curing at moderate temperatures, can be based on the catalytic curing agents K 61B and the imidazoles. Finally, the aromatic amines such as DDM and the anhydrides HPA and NMA are systems needing elevated temperature curing, but providing higher HDTs and hence maintenance of properties over a wider temperature range. Table 9.2 gives some typical properties of selected basic systems for small castings, etc.

The anhydrides are becoming widely used for all electrical casting applications and in particular the liquid anhydride NMA has shown considerable versatility in its use.[15] With NMA, the balance of properties of the cured casting can be varied over a wide range by alteration in the concentration of curing agent, the type and amount of catalyst employed, and the cure conditions. This is

Table 9.2 SOME PROPERTIES OF BASIC UNFILLED ELECTRICAL CASTING SYSTEMS[5]
(LIQUID DIGLYCIDYL ETHER OF DPP)

	Amine-ethylene oxide adduct	Imidazole	DDM	HPA
Initial mix viscosity;				
poise at 23 °C	70–80	100–140	60–90	10–13
poise at 60 °C	3–4	4–5	2–3	0·5–1·0
Pot life; 500 g, 23 °C	15–20 min	8–10 hr	7–8 hr	4–5 days
60 °C	—	15–20 min	1–1·5 hr	1·5–2·5 hr
Typical cure schedule	gel at 23 °C +1–2 hr at 100 °C	4 hr at 60 °C +2 hr at 150 °C	gel at 60 °C +4 hr at 150 °C	2–3 hr at 80 °C +4 hr at 150 °C
Heat deflection temperature, °C (ASTM D648; 264 lb/in^2)	45–95	85–130	145–150	125–130
Ultimate tensile strength, lb/in^2 (ASTM D638; 23 °C)	7 000	9–11 000	11 500	12 500
Ultimate flexural strength, lb/in^2 (ASTM D790; 23 °C)	16 000	17–19 000	17 000	18 000
Flexural modulus, lb/in$^2 \times 10^{-5}$ (23 °C)	5·25	5·6	4·0	3·8
Dielectric constant (ASTM D150; 50 Hz)				
23 °C	4·7	3·6	4·4	3·2
100 °C	—	3·7	4·4	3·4
Power factor (ASTM D150, 50 Hz)				
23 °C	0·008	0·004	0·004	0·003
100 °C	—	0·005	0·004	0·0015
Volume resistivity, ohm cm (ASTM D257; 23 °C)	1×10^{15}	5×10^{16}	1×10^{15}	2×10^{16}
Minimum dielectric strength) BS 2782 method 201A, V/10^{-3}in (23 °C)	350–400	350–400	375–425	350–400

partly illustrated in Figs. 9.3–9.6, which depict the change in HDT and 'crack rating', power factor, volume resistivity, and dielectric constant, at different temperatures with changing proportions of NMA.

The methods and equipment for encapsulating and potting are very similar to those for large casting manufacture, although usually

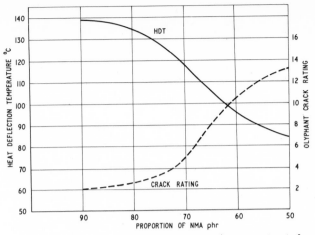

Fig. 9.3 Change in heat deflection temperature and Olyphant crack rating[5]

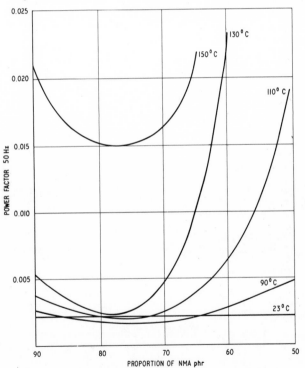

Fig. 9.4 Change in power factor at various temperatures[5]

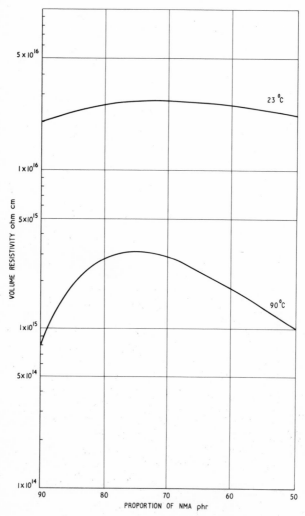

Fig. 9.5 Change in volume resistivity at 23°C and 90°C[5]

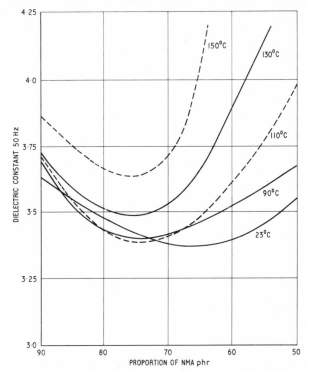

Fig. 9.6 Change in dielectric constant at various temperatures[5]

not so complex. Simple bench mixing and pouring methods accompanied by vacuum treatment of the separate ingredients and of the mixture is often used, but where large scale production is involved, as in the potting and encapsulation of electronic components, semi- or fully automatic machines are frequently employed which will meter, mix, and dispense two-component mixtures. These machines can incorporate vacuum degassing facilities and are able to dispense the resin mixture in single shots from 0·5 g to 5 kg or in a continuous stream up to 5 kg/min.

9.4.1 USES FOR SMALL CASTINGS, POTTING, AND ENCAPSULATION

Many different types of electrical equipment now use epoxide insulation and protection. The resins are used for transformer insulation where severe environmental conditions will be encountered, such as high temperatures and humidity; where mechanical shock will occur owing to sudden changes in electrical loading; or where a smaller power-to-weight ratio must be achieved. Electronic transformers (500 VA max) only incorporate epoxide resins when they are used in equipment for rocketry, airborne radar, etc. However, the smaller supply transformers up to about 5 kVA are now using an increasing amount of resin for their complete encapsulation in order to reduce size and weight. The resin system used is often highly extended or diluted, since the temperature and voltage requirements of these transformers are not high and a cheaper system can therefore be tolerated. Distribution transformers in the range 25 kVA to 5 MVA do not use epoxide resins to any extent at present in the U.K. Current practice is still oil immersion, whereas in the U.S.A. the smaller outdoor pole-mounted transformers are more and more being encapsulated to save weight. Similarly, very large power station transformers are not encapsulated in the U.K., in contrast again to the U.S.A.

The resin system used in all of these applications must have a sufficiently low viscosity and long pot life to ensure thorough impregnation of all windings before gelation occurs. This limits the amount of filler that can be used and may sometimes lead to the units first being impregnated with a very low viscosity system, followed by potting or encapsulation with a filled composition. Many other types of components and circuits have also been encapsulated successfully, including capacitors, resistors, transistors, chokes, relays. amplifiers, printed circuit and electronic circuit modules, and field coils of D.C. motors.[15-19]

9.5 IMPREGNATION, SEALING, AND DIP-COATING

Many electrical components are impregnated with epoxide resin insulation. Examples are small and large motor or transformer windings, capacitors, and solenoid coils. The units can be impregnated singly or in bulk, with or without encapsulation at the same time. The usual vacuum impregnation technique employed in the industry is used, together with a pressure cycle if necessary.

The resin system must possess a sufficiently low viscosity to allow complete penetration of the windings, yet not such that there is excessive run-off when the unit is withdrawn from the tank. In addition it must have a pot life sufficiently long at impregnating temperature to avoid gelling in the tank, yet be able to cure at moderate temperature fairly quickly. Liquid anhydride curing agents with a liquid diglycidyl ether resin are most commonly used, either NMA or eutectic anhydrides in conjunction with an accelerator. With the latter formulations, overnight cure at 100–120 °C is possible, which compares favourably with traditional phenolic-based varnishes. The NMA system, however, requires longer and higher temperature cure schedules. A very long pot life of the order of several weeks can be obtained with the anhydride mixture system by successive toppings-up of the resin mix with fresh material.

The insulation of electric motor windings is a good example of a different technique in applying epoxide resins. This method is called 'trickle impregnation'. In this process the units are dried and preheated to about 130 °C. They are then clamped at 15–30 degrees to the horizontal and rotated at 50–20 rev/min. The resin mixture is dripped on to the hot windings, and under the influence of gravity and capillary action flows down the windings in the slots and expels air along its path until all the coils are thoroughly impregnated along their entire length. The drip is then stopped, the armature rotated in a horizontal plane until the mix has gelled (usually twice the impregnating time), and the unit finally transferred to an oven for curing.

The advantages claimed for this impregnating process are:

(*a*) Complete void-free insulation without vacuum equipment.

(*b*) Shorter process time than the usual dipping and draining vacuum procedures, which are all eliminated.

(*c*) Reduction in material wastage.

The choice of the resin system is governed by the time required to impregnate the windings, and gel time should be twice the impregnation time. A liquid resin is most often used, polymerised with a liquid aromatic amine mixture. A useful review of the technique, the selection of the system, and the economics of the process is given by Adam.[20]

Resistors in one form or another are used in very large numbers throughout the electrical industry. The majority of the resistors are adequately insulated and protected by traditional materials such as phenolic resins. There is, however, a growing use of epoxide resins in specialist applications where improved properties such as moisture resistance are required, as in aircraft, radio and radar rocket-guidance systems, and computers. The resin system is usually applied by dipping the resistors into a tank of thixotropic coating material which has a long pot life and heat-stable thixotropy to prevent run-off during cure. Successful dipping systems are usually based on boron trifluoride-amine complexes as curing agents, with specific fillers providing the thixotropy.

Resistors can also be coated using a fluidised powder coating; alternatively, if severe mechanical shocks together with high humidities are encountered, they are completely encapsulated. Powder coatings are widely used for insulation purposes, for example sprayed powders insulating electric motor stators and rotors, or powders applied in a fluid bed being used to coat lamp ballasts and metal circuit boards, etc. Details of these powders are given in Chapter 8, Section 8.6.

Capacitors are found in all types of equipment,· and whilst materials such as polyesters, chlorinated diphenyls, waxes, etc. are used to impregnate the wound foil, the unit is often entirely coated with an epoxide formulation applied by dipping, or completely encapsulated. Certain capacitors are enclosed in aluminium or phenolic cases whose ends are often sealed with epoxide resins. For end-sealing the resin is heavily filled with sand or silica flour, and used with a low or medium temperature curing agent.

9.6 MOULDING POWDERS

An alternative method for encapsulating electronic circuits and small components is by the use of an epoxide resin moulding powder. This technique enables high speed mass production methods to be used in the on-line encapsulation of capacitors, resistors, commutative devices, and coils. Epoxide powders can be used in both compression and transfer moulding techniques[21] to fabricate articles such as bobbins, relay bases and connectors, silicon diodes, and metal film resistors.

Delmonte,[22] in considering epoxide resin transfer moulding powders, points out that PF and MF powders often require 1 000–5 000 psi moulding pressure, whereas epoxides require less than 1 000 psi and sometimes even as low as 100 psi. This is because of the better flow properties of epoxides, and enables critical electrical components to be encapsulated with the minimum of stress and distortion. Epoxide powders also flow at 100 °C and have HDTs greater than 200 °C, whereas other powders flow at temperatures greater than 150 °C.

Good flow properties at moderate temperatures and pressures also make the powders particularly suitable for moulding thin sections of material around relatively large metal inserts, such as a thin insulating coating round a stack of stator or rotor laminations. Low moulding pressures and viscosity also mean that forming around delicate pins and inserts presents no problem. In addition, the absence of volatiles being evolved during cure allows the production of large void-free mouldings such as bushings and insulators of large cross-section.

In general, moulding conditions are similar to those for other thermosetting powders, except in the case of transfer moulding, as mentioned above. The mouldings produced have excellent dimensional stability, low water absorption, and good resistance to tracking plus good electrical properties.

A similar method[13] of encapsulating, sealing, bonding, and protecting electrical equipment lies in the use of preformed pellets of B-stage resin of various shapes and sizes which merely need heat to melt them and allow the resin to flow into the mould or outer moulded case of the component. The resin then gels and cures to its final state. The advantages claimed for this method are the absence of capital outlay for plant and resin-handling equipment.

9.7 EPOXIDE RESIN LAMINATES IN ELECTRICAL APPLICATIONS

The preparation and properties of laminates based on epoxide resins are discussed fully in Chapter 10. This Section deals with their use in the electrical industry. For many years phenolic-resin-based laminates have been the standard rigid insulation material in electrical generators and motors, and their wide use will certainly continue where their performance is acceptable. However, in the search for laminates with improved properties, paper was joined by cotton as a reinforcing material, and melamine resins also began to be used as binders. Unfortunately, both paper and cotton laminates are hygroscopic, which leads to severe deterioration of electrical properties under high electrical stresses.

To improve water resistance and mechanical properties, glass cloth with phenolic or melamine resins was used, but these systems are not able to withstand the elevated temperatures that can develop in certain types of machine. In particular, phenolic board shows poor dielectric properties and bad tracking resistance at these temperatures.

Whilst silicone–glass laminates are now chosen for high temperature (class 180°C) insulation, epoxide–glass laminates are being used increasingly in the low and medium temperature range, i.e., up to class 150°C. Epoxide–glass laminates are lower in cost as compared with silicones, and have higher mechanical strength and moisture resistance. These advantages are also shown over phenolic resins, epoxides being superior in dielectric and tracking properties.

A major use for laminated material is in the manufacture of medium hp motors (up to 5 hp) and low power generators, the laminates being moulded as slot liners for the stators of the motors. In the larger motors (500 hp) the coil windings can be wrapped with an epoxide-impregnated mica tape, or alternatively dry mica tape can be used and the whole coil subsequently impregnated under vacuum with an epoxide resin system. Slot liners of epoxide resins are also used in these larger motors.

The laminates are usually made by the pre-preg process and for electrical use, reinforcing material is frequently glass cloth, mica paper, or mica splittings on glass cloth. The impregnated cloth is dried in a heated tunnel oven and the dried B-stage material is stored until required. To prepare the laminate, several layers of the B-stage stock are stacked in a heated press and consolidated under the influence of heat and pressure into one multi-ply laminate.

Mica, which is very widely used in high power equipment because its electrical, mechanical, and chemical properties are unaffected at high temperatures, can also be bonded with epoxide resins. Traditionally, mica splittings are bonded together with shellac or bitumen into a continuous sheet or tape, but the performance of this product is severely limited by the binder. Epoxide resin binders for mica provide a superior insulation material.

There are many epoxide systems that can be used for the production of slot liners and insulation sheet or tape for coil wrapping. One of the most popular formulations uses a liquid diglycidyl ether resin cured with DDM. This needs a short cure and exhibits good mechanical strength up to 155°C. Other curing agents are NMA, HPA, THPA, DDS, and boron trifluoride complexes. Improvements in high temperature performance of the laminates are also achieved by the use of an epoxidised novolak resin.

9.7.1 PRINTED CIRCUIT LAMINATES

Printed circuits are used extensively in all types of electronic equipment. They consist essentially of a laminated sheet made from a particular type of resin, reinforced with paper or fabric (usually glass cloth) on to one or both sides of which is bonded a layer of copper foil. The circuit is printed on the foil in acid-resisting ink and the excess copper etched off to leave a conductive circuit or circuits supported on a rigid sheet material. Holes are then punched in the sheet for fixing components such as transistors, resistors, and capacitors, which are soldered in place, usually by floating the copper-clad side downwards on a bath of molten solder, or by the wave-soldering technique.

The majority of printed circuit laminates used in domestic television and radio receivers and other non-critical industrial applications are made from paper-reinforced phenolic resins. These are moderately easy and cheap to make and there is therefore, at present, little potential use for epoxide-resin-based printed circuits in such outlets. For more critical applications, however, such as the closely packed circuits in computers, in complex telecommunications equipment, and in missile-guidance systems, materials of higher performance are required and epoxide-resin-based laminates have been found suitable.

In the U.S.A., the National Electrical Manufacturers Association (NEMA) publish specifications covering laminated sheet, rod, and tubes, including copper-clad laminates for printed circuits. NEMA specifications for epoxide resin copper-clad laminates have been introduced (G–10, G–11, FR–3, etc.), and these are also accepted as primary standards outside the U.S.A.

Both low-viscosity liquid and high-melting solid resins are used in printed circuit pre-preg manufacture. Solvents are used to control the handling viscosity of the formulation, and the degree of cure in the B-stage controls the melt-viscosity of the system. In particular, the liquid diglycidyl ether (WPE 180–200) and solid resin (WPE 450–500) are the most widely used, together with a brominated resin for flame-retardance or an epoxidised novolak for improved laminate resistance to degreasing or solder baths, when required. Common curing agents include DICY, DDS, BF_3–MEA, and BDMA. For the general purpose glass-fabric-based G–10 laminate, the solid resin cured with DICY is mostly used. For the higher temperature resistant board (G–11 grade) curing agent DDS accelerated with BF_3–MEA is used to polymerise the liquid resin. Further details of the G–10 grade are given in Chapter 10, Section 10.3.3.

9.7.2 LAMINATES IN SWITCHGEAR APPLICATIONS

A large amount of epoxide resin is consumed in the manufacture of switchgear equipment where laminates are used for mechanical linkages, arc quenching chambers, and, most important of all, for air-blast circuit breaker tubes. Circuit breakers are safety devices which operate when the current in a circuit rises rapidly because of a fault or breakdown in the system. Current transformers monitor each line and set off the circuit breaker to isolate the section of line when sudden rises in current occur. Heavy duty circuit breakers extinguish the arc formed when the contacts part by blowing it out with a very powerful blast of air at high pressure. This air blast is rather like a shock wave, and the tube along which the air is directed must therefore be very strong and also a good insulator. Epoxide–glass laminates fulfil these two requirements. The tubes can be made by a filament-winding technique, although another method is to wrap a specially made glass cloth round a mandrel which fits into a horizontal heated mould. The resin mixture is then allowed into the mould under vacuum, where it thoroughly impregnates the glass cloth. The mould is heated to cure the resin and the excess of hardened resin is subsequently removed by machining.

9.8 EPOXIDE RESIN CABLE-JOINTING SYSTEMS

Resin cable-jointing kits are now becoming widely used as a replacement for the traditional bituminous product. The bitumen compounds require melting and hot pouring on site, with subsequent topping up. The epoxide resin compounds require no heat for making the joint, have low shrinkage on cure, and exhibit no creep or flow during service. In addition, simple and inexpensive moulds can be used and fast-setting formulations reduce the time to make a joint.

REFERENCES

1. ANON.,'Survey of Electrical Insulating Materials', *Electl. Rev., Lond.,* **174,** No. 5 (1964)
2. ANON., *Electl. Times,* **148,** No. 4, 126 (1965)
3. DUNN, P. A., *Electl. Times,* **153,** No. 26, 19 (1968)
4. PARRY, H. L. and MACKAY, H. A., *S.P.E. Jl.,* **14,** No. 7, 22 (1958)
5. Shell Chemical Co., Technical Literature
6. ROTHWELL, K., *Electl. Rev., Lond.,* **177,** No. 3, 89 (1965)

7. DORMAN, E. N., *Proc. 6th Electl. Insulation Conf., U.S.A.,* paper IEEC 32C3–2 (1966)
8. YARSLEY, V. E., GRANT, W. J. and IVES, G. C., *Electl. Res. Ass., U.K.,* Tech. Rep. L/T 274
9. ASH, D. O. and DEY, P., *Electl. Times,* **141,** No. 2, 41 (1962)
10. KESSEL, A. A. and NORMAN, R. S., U.S. Pat. 2,997,527
11. BILLINGS, M. J. and HUMPHREYS, K. W., *Proc. 6th Electl. Insulation Conf., U.S.A.,* paper 32C79–71 (1966)
12. PATRICK, C. T. and MCGARY, C. W., *S.P.E. Ann. Tech. Conf.,* (1966)
13. Ciba (ARL) Ltd., Technical Literature
14. WEBER, W. E., *S.P.E. Jl.,* **14,** No. 3, 49 (1958)
15. MUELLER, B. H. and HARPER, C. A., *Electl. Mfr.,* **65,** 2, 119 (1960)
16. HONSINGER, V. B., *Electl. Mfr.,* **64,** 5, 114 (1959)
17. CRAWFORD, D. E., *Mat. in Design Engng.,* **46,** No. 1, 99 (1957)
18. JEWELL, J. M. and JENNER, W. C., *Mod. Plast.,* **37,** No. 4, 101 (1959)
19. ANON, *Electl. Times,* **153,** 23, 929 (1968)
20. ADAM, H., *Kunststoffe,* **54,** 490 (1964)
21. ZECHER, R. F., *Insulation, Lake Forest,* **13,** No. 1, 37 (1967)
22. DELMONTE, J., *Insulation, Lake Forest,* **13,** No. 2, 69 (1967)

BIBLIOGRAPHY

Insulating Materials for Design and Engineering Practice, CLARK, F. M., Wiley, New York (1962)
Electrical Encapsulation, VOLK, M. C., LEFFORGE, J. W. and STETSON, R., Reinhold, New York (1962)

CHAPTER 10

Laminates

10.1 INTRODUCTION

The term laminate, or composite, is used here to describe fibre-reinforced polymers, and unless otherwise stated it is assumed that the reinforcing material is glass fibre in one of its many forms.

The earliest use of fibrous reinforcing materials with polymers was the use of cotton and paper to improve the strength properties of phenolic resins. Then in the early 1940s the need for a suitable material to construct aircraft radomes led to the introduction of glass-fibre-reinforced polyester resins. Since that time there has been a continuing development of reinforced plastics technology embracing different types of resin system, reinforcements, and methods of laminate fabrication, but in general glass has held its position as the most important reinforcing material and polyester resins are used in the great majority of applications.

Laminates are made up of two different materials with contrasting strengths and elasticity. Glass fibres can show a Young's modulus of about 10×10^6 psi, ultimate tensile strengths in practice of around 200 000 psi and a linear stress/strain relationship to failure with no yield. Unreinforced resin systems by comparison have ultimate tensile strengths of from 5–15 000 psi, moduli of 0·3 to $0·5 \times 10^6$ psi and a non-linear stress/strain relationship. Combining the two materials, it is possible to produce an epoxide resin composite with an ultimate tensile strength of about 60 000 psi and modulus $4·0 \times 10^6$ psi. If the composite is based on unidirectional glass fibre, then tensile strengths of about 200 000 psi can be achieved with a modulus of 6–8 $\times 10^6$ psi.

Thus the orientation of the reinforcing fibres is an important factor in determining the properties of the laminate, unidirectional, bidirectional, and random orientation being possible. Reinforced plastics do not therefore show the same properties in all directions, i.e., they are anisotropic; even random-orientated fibres give uniform properties only in the plane of the reinforcement. Pickthall[1] gives ample illustration of this point and draws up some general rules on the effect of fibre orientation on mechanical properties of composites.

Since the reinforcement in a composite is the chief source of strength, the characteristics of the cured resin matrix are extremely important in transmitting the applied stresses to each fibre. It is therefore apparent that the critical point in the material is the interface between resin matrix and fibre, and poor resin-to-glass adhesion can lead to rapid failure of the composite when stressed. Not only is the adhesive nature of the resin system important but the chemical, electrical, and thermal stability characteristics are also factors which affect the ultimate performance of the laminate.

The relative amounts of resin and fibre reinforcement also play a part in determining strength properties. In general, increasing resin content leads to decreasing strengths with the formation of resin-rich areas. Laminates produced by contact pressure have typical resin contents of up to 60–70% by volume, although the actual level depends very much on the viscosity of the resin mix and the skill of the individual operator. The use of pressure laminating helps to reduce resin content and hence improves strength properties.

The more expensive epoxides are used only in applications where their performance advantages over polyesters are fully justified. These advantages are:

(*a*) excellent adhesion to reinforcements,
(*b*) low shrinkage during cure,
(*c*) good mechanical and electrical properties,
(*d*) high heat-resistance,
(*e*) good fatigue-resistance, and
(*f*) good chemical- and moisture-resistance.

It is interesting to compare the typical strengths of epoxide laminates with other materials. Table 10.1 indicates the high strength-to-weight ratio of epoxide resin laminates as compared with steel and aluminium, but also shows their low modulus of elasticity. The fatigue strength of epoxide laminates is also superior to other glass-reinforced plastics. Pickthall[1] quotes typical values for the endurance limit at 10^6 c/sec when expressed as a percentage

Table 10.1 STRENGTH PROPERTIES OF EPOXIDE RESIN LAMINATES COMPARED WITH
STEEL AND ALUMINIUM

Material	Specific Gravity	Tensile Strength, lb/in² ($\times 10^3$)	Young's modulus, lb/in² ($\times 10^6$)
Structural steel (low carbon)	7·8	50–70	29·0
Aluminium	2·7	10–35	10·0
Filament wound epoxide/glass laminate	2·0	80–250	4–8·0
Flat epoxide/glass cloth laminate	1·8	60–80	2–3·5
Flat polyester/glass cloth laminate	1·7	25–50	2–3·0

of the initial strengths:

	%
Polyester/glass	23
Epoxide/glass	32
Aluminium alloys	12
Titanium	8

Other long-term properties under continuous loading, such as
creep, have been considered by Thompson,[2] and extensive data on
the heat-ageing of epoxide laminates are also available.[3]

10.2 FIBROUS REINFORCEMENTS

There are very many different types of fibrous reinforcement that
can produce a composite structure with epoxide resins. The synthetic fibres nylon, Terylene, and the acrylics are all suitable, and
form highly flexible laminates with good abrasion and chemical
resistance. However, they are almost always used in conjunction
with glass fibre reinforcements to provide special properties. Other
materials such as cotton, hessian, mica, and paper may also be used,
but the laminates produced have properties inferior to those of
laminates based on glass fibre. Paper has a particular use as the
reinforcement for certain electrical laminated board for printed
circuits. The paper-based laminate is prepared by high-pressure
hot-moulding of the paper pre-impregnated with the resin system.

The most important reinforcing material for epoxide resin laminates is glass fibre, since it offers the best balance of properties. In
particular, the high tensile and compressive strengths and (in certain
types of unidirectional cloth) the high modulus of elasticity that can
be obtained, coupled with good dimensional stability and relative

inertness to many chemicals, are all factors that contribute to the wide use of glass.

Two types of glass are generally available, E or electrical grade, and A or alkali grade. E glass was originally developed for electrical insulation and is a borosilicate glass with very low alkali content. Thus the action of moisture or alkali on E glass does not lead to the leaching-out of soluble salts, though it is attacked by acids. In addition, it has 10% greater tensile strength than A glass, a high-alkali glass which loses soluble salts by leaching under neutral or alkaline conditions. In general, the stronger E glass is used with epoxide resins, though with adequate covering by resin A glass and E glass laminates are both satisfactory.

Glass in the form of fibres has a very much greater strength than ordinary bulk glass, probably because of the elimination of surface flaws. For E glass, measurement[4] of the ultimate tensile strength has achieved 500 000 psi for fibres of diameter 0·00020–0·00060 in on laboratory-produced samples. Under production conditions it is usual to obtain fibres with tensile strengths within the range 180–220 000 psi. The drop in strength is probably due to flaws being caused by damage to the fibre.

Glass fibres are produced in a continuous drawing process, the filaments being sized to bind groups of them into a strand. The strands are twisted into yarns, and the yarns woven into cloth. A wide combination of weaves, weights, and finishes is available, but one of the strongest fabrics, and the one most widely used with epoxide resins, is a satin-weave cloth. Alternatively, the strands are used unchanged, as in the filament-winding process, or chopped into short lengths to form chopped strand mat. Strands can also be collected into bundles, held together by a binder but not twisted. These are called rovings.

After these processes have been completed the organic size on the filaments is removed, usually by heating. The heat-cleaned glass may then be used without further treatment and its adhesion with resins is excellent. However, the glass surface is easily altered by the ready absorption of a film of water from atmospheric humidity. It is therefore usual to apply an organic finish to the glass to protect against the moisture film and to act as a coupling agent between glass fibre and resin matrix. The way in which the finish improves laminate strength is not clear, although the finishes have been designed to contain functional groupings that will bond both to the glass and to the resin molecule. The usual finishes for epoxide resin laminates are based on a chromium acrylate complex, or a silane containing amine or epoxide groups. Lee and Neville[5] review the various available surface treatments and their possible modes of action.

For a full description of the manufacture of glass fibre, and the various types of rovings, weaves, and finishes, the reader is referred to some of the more comprehensive texts on the subject.[6-8]

Recently, much attention has been paid to the development of resin–carbon fibre composites: and epoxide resins are at present the most satisfactory in this field. Phillips[9] has reported on two types of fibre available, a high modulus grade designated Type I and a high strength grade designated Type II. Type I has a specific gravity of 2·0, ultimate tensile strength $250–300 \times 10^3$ psi, and Young's modulus $55–60 \times 10^6$ psi. Type II has a specific gravity of 1·7, ultimate tensile strength $400–450 \times 10^3$ psi, and Young's modulus $35–40 \times 10^6$ psi. The Young's moduli possessed by these highly oriented polycrystalline fibres will therefore enable composites to be made with very high specific stiffness.

10.3 LAMINATING PROCESSES, SYSTEMS, AND PROPERTIES

All the processes produce a composite structure consisting of consecutive layers of fibrous reinforcing material thoroughly impregnated throughout all layers by the resin. On curing the resin the two-phase composite material is formed. The techniques used fall into two categories, wet lay-up and dry lay-up. The former involves the use of a liquid resin system to impregnate the reinforcing material either before or after it has been placed on the mould or mandrel. The laminate is then formed under contact or at very low pressure, or at a predetermined tension in the case of filament winding. Alternatively, high pressure matched-metal die-moulding can be carried out.

Dry moulding is essentially a high-pressure process, and is identical with the conventional laminating technique used for phenolic, melamine, and urea resins. In these cases the reinforcing material in sheet or tape form is impregnated with the resin system in a carrier solvent. The solvent is then removed, leaving a dry resin-impregnated reinforcement which is moulded under heat and pressure using matched dies or used in a filament-winding process.

It is not intended to consider in detail the various production techniques for laminates. They have been described in many publications, often in relation to polyester technology, though of general applicability.[6–8] Instead, a few of the more widely used resin formulations and the properties of their laminates are considered, together with a brief indication of their applications.

10.3.1 WET LAY-UP METHODS

The hand lay-up technique, usually associated with curing at contact pressure only, is the simplest of all. It requires the minimum of equipment and is used to produce large complex structures not suitable for a pressure process, such as tanks and drilling jigs. It is also used for articles required in small quantities where the number produced does not justify large investment in tooling costs. Laminates made in this way do not show the best performance that epoxides can provide, since the resin content is usually high (about 60% by volume). Moreover, the process is a manual one, the quality of the laminate depending to a large extent on the skill of the individual moulder. Either male or female moulds made from wood, metal, plaster, etc., can be used and their design can be extremely simple. The mould surface is first treated with a release agent, which is followed by a face (or gel) coat of filled resin mix. When this coat has attained a tacky but not fully hardened state, a layer of glass cloth is laid on it and thoroughly impregnated with unfilled resin mixture. The laminate is then built up by applying successive layers of reinforcement, each being impregnated with the resin, often by using a stiff brush. Entrapped air and excess resin are removed by rolling with a split-washer roller, and the laminate allowed to harden, usually at room temperature. Heat can be applied at the curing stage by various means such as portable electric heaters or hot air blowers, which greatly assists the rate of cure and enhances the physical properties of the laminate.

The process as described is a very simple one, requiring only hand tools and perhaps hand-mixing of the resin system. It can be made more sophisticated by:

(*a*) applying the resin mixture to the glass cloth via a spray gun, or

(*b*) spraying both resin mix and chopped glass fibre on to the mould surface either from two separate guns or one composite gun.

In both cases the deposited materials must be rolled to consolidate the laminate.

Table 10.2 EPOXIDE RESIN SYSTEMS USED IN CONTACT PRESSURE LAMINATING, AND THE STRENGTHS OF LAMINATES BASED ON THEM

System	Resin content (%)	Ultimate flexural strength lb/in² (×10³)	Modulus lb/in² (×10⁶)
Liquid diglycidyl ether resin + DTA (10 phr)	34	52	2·0
Diluted resin + DTA (10 phr)	33	35	1·7
Diluted resin + TET (12 phr)	29	50	3·0
Diluted resin + ethylene oxide adduct of DTA (25 phr)	33	62	2·3
Liquid resin + amine adduct (25 phr)	35	67	2·8

Note : (1) All laminates based on glass cloth and cured for 7 days at 25°C
(2) Diluted resin contains n-butyl glycidyl ether

Suitable resin systems for use in the contact pressure method are given in Table 10.2, together with physical properties of laminates based on them. These are all room-temperature curing systems, although any heating even to 40–50°C will improve ultimate properties.

Whilst not exhibiting the very high strength properties of other epoxide laminates, these systems are widely used in tooling and model making, where their dimensional stability, light weight, and high strength-to-weight ratio offer advantages over conventional materials. Drilling jigs, checking fixtures, Keller models, and moulds for polyester laminating are all established uses. In addition there are many miscellaneous uses such as radomes, chimney stack linings, tank lining and repair, lorry cabs, wooden boat-hull sheathing and repair, and concrete moulds.

Where better quality laminates are required, and the shape, size, and economics of the job permit, it is possible to apply low pressure to the laminate during the curing process. This can be achieved by the vacuum bag method. This usually consists of laying-up the laminate as described previously and then placing a flexible bag (perhaps of PVA) over the mould, which is clamped down round its circumference. A vacuum is then drawn under the sheet and atmospheric pressure forces air and excess resin out of the laminate. This can be assisted by rolling the laminate while the vacuum is being applied. For moulding at higher pressures, the normal matched-metal die-moulding method is quite suitable.

A novel wet lay-up method for the large-scale production of sizeable mouldings such as boats and radomes is the resin injection technique. The required amount of glass reinforcement is placed between matched moulds suitably treated with release agent. Resin and curing agent are then mixed and pumped through the mould under pressure, removing all entrapped air and impregnating the glass cloth; and the laminate is cured by using heaters built into the surfaces of the mould.

All wet lay-up techniques are suitable for high-temperature curing systems, the limitation being the ability to reach and maintain the optimum cure temperature for large hand lay-up mouldings. However, the main reason why these systems are not used in the contact pressure process is that the high resin content obtained by this method would not allow the full strength properties of the temperature curing systems to be developed. Thus these latter systems are almost always associated with a heat and pressure process which would include filament winding.

Table 10.3 STRENGTH PROPERTIES OF LAMINATES BASED ON HOT CURING EPOXIDE RESIN SYSTEMS

System*	Resin content (%)	Care schedule (hr/degC)	Flexural strength (lb/in²)			Modulus (lb/in² × 10⁶)		Ultimate tensile strength (lb/in²; at 25°C)	Tensile modulus (lb/in² × 10⁶; at 25°C)
			25°C	120°C	150°C	25°C	120°C		
MPD (14 phr) or DDM (27 phr)	30	1/100 +1/200	84 000	65 000	48 000	3·7	3·3	54 000	3·4
HPA (80 phr)	25	1/100 +1/200	91 000	15 000	—	3·8	1·6	58 000	4·0
NMA (90 phr)	24	1/100 +15/200	79 000	—	41 000	3·4	2·5	61 000	4·2
DDS (20 phr) with polyphenol glycidyl ether resin	33	1/130 +2/220	68 000	—	46 000 (at 200°C)	—	—	—	—

*Resin used: low molecular weight liquid diglycidyl ether

Some resin systems that require elevated temperature curing conditions are shown in Table 10.3 together with the strength properties of their laminates, and typical chemical resistance properties for laminates based on MPD as a curing agent are listed in Table 10.4.

Table 10.4 PERCENTAGE LOSS IN FLEXURAL STRENGTH OF MPD LAMINATES AFTER IMMERSION FOR ONE YEAR IN A NUMBER OF DIFFERENT REAGENTS AT $25\,^\circ\text{C}$[3]

Reagents	*% Loss in strength*
Sulphuric acid 95%	Delaminated after 7 days
Sulphuric acid 70%	2
Sulphuric acid 3%	34
Hydrochloric acid 37%	45*
Hydrochloric acid 10%	32
Nitric acid 50%	Top layer delaminated
Nitric acid 30%	91*
Phosphoric acid 98%	9
Phosphoric acid 10%	9
Acetic acid, glacial	11
Acetic acid 10%	16
Oxalic acid, saturated	41*
Sodium hydroxide 50%	9*
Sodium hydroxide 1%	13
Ammonium hydroxide 28%	67
Sodium chloride 20%	6
Sodium sulphate 30%	9
Ammonium nitrate 50%	24
Acetone	28*
Ethyl acetate	11
Ethylene dichloride	10
Ethylene glycol	4
Hydraulic brake fluid	11
Jet fuel	2
100 Octane gasoline	9
Methyl ethyl ketone	1
Isopropyl alcohol	11
Allyl chloride	6
Hydrogen peroxide	Delaminated
Water, distilled	2

*After 180 days

Uses for these laminates are widespread, occuring in the aerospace, automobile, electrical, and engineering industries.

10.3.2 FILAMENT WINDING

This is a special type of wet lay-up laminating, and is probably the most important fabricating technique for structural epoxide laminates. It consists of impregnating continuous glass roving or tape (or other reinforcement) with the binder resin and then winding it on to a mandrel having the shape of the final product, in one of many winding patterns. The glass rovings are kept in tension throughout the winding process. In addition, the pattern is carefully calculated and controlled to give the required hoop and longitudinal strengths and to orient the filaments so that in service they are stressed in tension, as far as possible. After winding, the laminate is usually hot-cured in an oven and then removed from the mandrel, though in some cases the mandrel remains part of the finished article.

Dry filament winding is a recent development in the U.S.A.[10] This consists of using pre-impregnated roving or tape which is made in a similar way to pre-preg cloth. The dry tape is then wound on the mandrel, followed by oven curing.

Filament winding can produce cylinders from $\frac{1}{8}$ in to 20 ft in diameter. Any shape can be made if it can be wound under tension and if it has an axis of rotation.

Starting from equipment used to produce light-weight high-strength chemically resistant pipe, considerable development work has led to fully automatic winding machinery, controlled by punched card or tape programmes which specify the exact winding pattern to be carried out. This equipment is now producing pipe,[11] chemical storage tanks, railway tanks, cars, sporting goods, insulating rod, air-blast circuit breaker tubes, and many other items. In addition the U.S. space programme[13] and general military requirement has used filament-wound epoxide laminates very widely for rocket motor cases, rocket launchers, pressure bottles, space satellite walls, torpedo and mine cases, and deep submergence vessels. (Flat sheet is also made by splitting a filament-wound cylinder along its longitudinal axis and flattening out the curved piece.)

The major advantages of the filament winding process are:

(*a*) it produces very high strength-to-weight ratio composites,
(*b*) it can be made fully automatic, and
(*c*) the composites can be made to closely controlled reproducible dimensions and performance properties.

An indication of the very high strengths that can be obtained for filament-wound epoxide glass composites is given by Riley *et al.*[12] (Table 10.5).

Table 10.5 TYPICAL·PROPERTIES OF FILAMENT WOUND EPOXIDE/GLASS LAMINATES[12]

Property	Type of winding	
	Circumferential	Helical
Density, lb/in^3	0·075	0·075
Ultimate hoop strength, lb/in^2 ($\times 10^3$)	180–230	120–135
Tensile modulus of elasticity, lb/in^2 ($\times 10^6$)	6–8	3–5
Flexural strength, lb/in^2 ($\times 10^3$)	280	100
Interlaminar shear strength, lb/in^2 ($\times 10^3$):		
Parallel to fibres	3–6	3–6
Perpendicular to fibres	18–20	18–20
Compressive strength, lb/in^2 ($\times 10^3$)	70	70
Tensile strength/weight ratio ($\times 10^6$)	2–2·5	1·6–1·8

Further information on all aspects of the filament winding process is given by Rosato and Groves,[14] and in various papers presented at the London Conference on Filament Winding, October 1967.[15]

Many different resin systems have been developed for the wet-winding process, and all have sufficiently low viscosities for good wetting of the glass, long pot life to allow large structures to be wound, and modest temperature cures to keep costs down and allow winding over heat-sensitive materials. Table 10.6 gives the properties of two widely used systems. The strength values were determined by NOL (U.S. Naval Ordinance Laboratory) ring tests, the rings being cut from cylinders 2 in long.

10.3.3 DRY LAY-UP METHODS

One limitation to the wet lay-up method is that it is restricted to resin systems which have low viscosities at working temperatures. Attempts to overcome this led to the use of reactive diluents, which unfortunately often degrade the properties of the final laminate, or to working at higher temperatures which causes a reduction in pot life, sometimes to an impractically short period. One advantage of the dry lay-up method is that it does not suffer from this viscosity limitation, and higher molecular weight solid resins can thus be used.·

The technique involves impregnation of the reinforcing cloth or rovings with the resin system, usually dissolved in a suitable solvent such as acetone or methyl ethyl ketone or possibly a glycol ether, although some liquid resin systems can be used without

Table 10.6 STRENGTH PROPERTIES OF FILAMENT WOUND LAMINATES BASED ON TWO
TYPICAL EPOXIDE RESIN FORMULATIONS[3] DETERMINED BY NOL RING TEST*

	Liquid diglycidyl ether resin + DDM (27 phr)	Liquid diglycidyl ether resin + NMA (90 phr) + K.54 (1 phr)
Hoop tensile strength, lb/in²		
23°C	215 000	170 000
150°C	190 000	174 000
Tensile modulus		
23°C	$6·3 \times 10^5$	$5·7 \times 10^5$
150°C	$6·2 \times 10^5$	$5·8 \times 10^5$
Flexural strength, lb/in²		
23°C	173 000	190 000
150°C	150 000	140 000
Flexural modulus		
23°C	$7·0 \times 10^5$	$7·3 \times 10^5$
150°C	$6·9 \times 10^5$	$6·8 \times 10^5$
Horizontal shear strength, lb/in²		
23°C	9 600	8 600
150°C	6 200	5 100
Resin content	15·0%	14·8%
Cure schedule	3 hr at 150°C	2 hr at 125°C
		2 hr at 200°C
		2 hr at 250°C

*The NOL rings were cut from 2 in long cylinders wound with 12-end E–HTS
(Owens Corning high tensile glass finish) and winding carried out at 40 rev/min
using 10 lb of total tension on the rovings

solvents. The solvent is then removed in a drying oven and the
impregnated stock is ready for laminating, usually in a heated press
with matched dies.

Both liquid and solid diglycidyl ether resins and epoxidised
novolaks can be used in conjunction with a wide range of curing
agents. Aromatic amines such as MPD and DDM are particularly
suitable since in these systems cure can be partially advanced to a
point (the B-stage) where the system solidifies, and after this stage
cure proceeds only slowly at room temperature. In practice, liquid
resin and aromatic amine are dissolved in the solvent at room
temperature to a 50% solids concentration. The reinforcing
fabric is then impregnated with the solution by hand or mechanical
treater and the fabric dried overnight at room temperature or for
20 min at 90°C in a drying oven. During this period the system
reaches the B stage of cure. The resulting pre-preg material has a
maximum shelf life of 30 days at 25°C. It can then be cured in a

press at 25–200 lb/in^2 and 140–150°C for 30 min, followed by a post-cure of 2 hr at 150°C.

Pre-pregs with longer shelf lives can be obtained by using curing agents such as DICY or BF$_3$ adducts. These are latent curing agents and do not react with resin at room temperature. DICY (3–4 phr) is often used in conjunction with a low melting point solid resin together with a small amount of an accelerator such as BDMA (0·2 phr), the system being dissolved in a glycol ether. Pre-preg cloth from this mixture has a shelf life of up to one year and is a dry rigid board. It is best used, therefore, for the preparation of flat laminates. In general, pre-preg cloth based on the solid resins is stiff and cannot be draped easily over a contoured surface. Drape characteristics can be improved by blending a liquid resin into the solid resin system. Acid anhydrides such as HPA, NMA, and HET are also used to prepare dry lay-up systems. However the cure accelerator normally used with them precludes the possibility of B staging, and the impregnated stock must be stored under refrigerated conditions.

Laminating with pre-impregnated reinforcement is a method of controlling the important variable of glass content and hence ultimate quality of the laminate. The most widespread use of this technique is in the U.S.A., for the manufacture of laminated board for electrical use, frequently as printed circuit base boards (see Chapter 9.7.1).

The NEMA G–10 grade laminate, for example, has a high mechanical strength (flexural, impact, and bonding) at room temperature and in addition good dielectric and electrical strength properties under both dry and humid conditions. A resin system that meets these requirements is the low molecular weight solid glycidyl ether resin (WPE 450–500) cured with DICY (4 phr) plus BDMA. Glass cloth, impregnated with this solution and dried, has a useful shelf life of 6 months. On curing 10 plies of the cloth in a press at 500 lb/in^2 for 30 min at 170°C, a laminate of 35% resin content was obtained[16] which had a flexural strength of 57 900 lb/in^2 and flexural modulus $3·3 \times 10^6$. Other strength properties were compressive 27 900 lb/in^2; tensile 36 300 lb/in^2; tensile modulus $2·5 \times 10^6$; and ultimate elongation 1·8%.

Composites of epoxide resins and carbon fibres have been prepared by the pre-preg method using a 'leaky mould' technique[9] to allow removal of the surplus resin after the fibres are wetted. Some typical test results of a unidirectional composite made in this way are given in Table 10.7. The resin system used was a liquid diglycidyl ether cured with 20 phr of DDS plus 1 phr of BF$_3$–400.

Table 10.7 TYPICAL MECHANICAL PROPERTIES OF CARBON FIBRE COMPOSITES[3]

Fibre volume fraction and type		Flexural strength (lb/in²)	Flexural modulus (lb/in² × 10⁶)	ILSS (lb/in²)
0·4–0·5	It	113 000	12·6	9·200
	Iv	73 000	14·5	3 500
	IIt	131 000	9·5	> 11 300
	IIv	115 400	9·0	9 400
0·5–0·6	It	118 400	15·0	> 8 500
	Iv	69 000	12·2	3 300
	IIt	160 000	11·4	> 11 200
	IIv	142 400	11·4	9 300
0·6–0·7	It	102 000	16·4	> 8 800
	IIv	131 000	12·4	7 700
0·7–0·8	IIv	136 200	12·5	6 800
	IIt	205 300	13·8	11 900

Note

(a) Fibre type Iv = Type I virgin
    Fibre type It = Type I treated

(b) Flexural strength measured by the ASTM D790–63 method using a crosshead speed of 0·05 in/min and a test specimen of span-to-thickness ratio 16:1.

(c) ILSS = Interlaminar shear strength, determined by the ASTM D2344–6ST method. The symbol > indicates that the specimen broke in tension at the outside surface instead of in shear.

(d) The composite was prepared from pre-impregnated fibre laid unidirectionally in the mould, the latter then being placed in a press at 170 °C for 12 min. After this time the press was slowly closed to stops over 1 min. and the mould allowed to remain for 1 hr at 170 °C and 300 lb/in².

REFERENCES

1. PICKTHALL, D., *Engng Mater. Des.*, **6**, No. 6, 408 (1963)
2. THOMPSON, A. W., *Reinfd. Plast.*, **1**, No. 11, 4 (1957)
3. Shell Chemical Co., Technical Literature
4. THOMAS, J., *Nature*, **181**, 1006 (1958)
5. LEE, H. and NEVILLE, K., *Handbook of Epoxy Resins*, McGraw-Hill, New York 22, 17 (1967)
6. DE DANI, A., *Glass Fibre Reinforced Plastics*, Newnes, London (1960)
7. MORGAN, P. (Ed.), *Glass Reinforced Plastics*, Iliffe, London (1961)
8. DUFFIN, D. J. and NERZIG, C., *Laminated Plastics*, Reinhold, New York (1958)
9. PHILLIPS, L. N., *Chemy. Ind.*, No. 17, 526 (1968)
10. CHRISTIE, S. H., *4th Int. Reinforced Plastics Conf. of Brit. Plastics Federation*, paper 15 (1964)
11. Bristol Aeroplane Plastics Ltd., Technical Literature
12. RILEY, M. W., *et al.*, 'Filament Wound Reinforced Plastics', *Special Report of Mat. in Design Engng.*, **52**, No. 2, 127 (1960)
13. LUBIN, G. and ROSATO, D. V., *4th Int. Reinforced Plastics Conf. of Brit. Plastics Federation*, paper 35 (1964)

14. ROSATO, D. V. and GROVES, Jnr., C. S., *Filament Winding*, Interscience, New York (1964)
15. Plast. Inst. Lond., Conf. on Filament Winding, reported in *Plast. and Polym.*, **36,** 122 (1968)
16. Dow Chemical Co., Technical Literature

CHAPTER 11

Adhesives

11.1 INTRODUCTION

The adhesive bonding of materials is now well established in many industries and offers a number of advantages over conventional joining methods. Adhesive bonding avoids the stress concentrations set up when rivets, bolts, or welds are used to join materials, the stresses in a correctly designed adhesive joint being spread evenly throughout the glue line. The use of an adhesive can also overcome the possibility of galvanic corrosion between dissimilar metals and eliminate the need for drilling materials. Hence, joints can be formed that are made vacuum- or pressure-tight more easily.

Adhesives based on epoxide resins have outstanding strength properties, powerful bonds being formed between similar and dissimilar materials such as metals, glass, ceramics, wood, cloth, and many types of plastics. In fact, the cohesive strength of a properly cured resin within the glue line is also very high so that sometimes failure under stress occurs within the material being bonded rather than within the glue line itself (e.g., an aluminium-to-aluminium bond).

The variations in epoxide formulation ensure that a range of adhesives of very wide versatility can be prepared. This has allowed the applications for epoxide resin adhesives to develop from their original use for metal bonding in the aircraft industry to widely differing industries. In construction work, epoxide-polysulphide adhesives join old concrete to new. Precast concrete beams and blocks are bonded together, and steel fitments bonded to dry or damp concrete, also using this type of formulation.[1] The adhesives have even been used to bond a steel sleeve joining two sections of reinforced concrete piles, the bond easily withstanding the subsequent pile driving. Pipes carrying corrosive effluents have been bonded with epoxide adhesives,[2] utilising their outstanding chemical resistance. Other constructional uses include pile bonding, jointing in aluminium window frames, bonding railway line base-plates to sleepers and traffic markers to concrete roads. An important and widespread use[3] is the repair of cracks in concrete. The adhesive is either poured, brushed, or pumped under pressure into the cracks, and this technique has been used successfully to repair dams, bridges, breakwaters, paths, and roads (see also Chapter 12, Section 12.3).

An unusual and widely accepted technique for making stained glass windows is to use an epoxide adhesive in place of the traditional lead. The concrete ribs and panels of the 75-ft lantern of the new Liverpool Cathedral are all bonded together with over a ton of an epoxide resin adhesive.

These adhesives have a wide variety of applications in the aircraft industry where high performance is essential. One important use is the construction of thin-skinned honeycomb panels where the epoxide adhesive is used to bond the aluminium skin facing to the honeycomb. These panels have high strength-to-weight ratios together with good thermal and vibrational resistance, and are used for such things as aircraft fusilage doors, side panels, floors, elevators, rudders, and trim tabs.

The electrical industries are very large consumers of epoxide resins, some of which are used primarily for adhesive bonding of items such as the shafts of gramophone motor rotors, metal laminations of motor stators, and porcelain insulators. Epoxide adhesives are also used in bonding the bristles to the handles of paint brushes and the grit to the metal in grinding wheels, in domestic adhesives, in the repair of ceramics, and in numerous other applications.

The adhesives are most commonly used as two-component fluids or pastes, which cure at room or elevated temperatures. Some adhesives are supplied in rolls as a supported sheet, i.e., a glass fabric tape impregnated with adhesive, which is simply called a tape. Polythene on both sides of the tape keeps it from sticking to itself in the roll. Just before use, the polythene is stripped off and discarded. Other forms of the adhesive are powders, pellets, or rods. The tapes and various forms of solid adhesive contain both resin and curing agent component. There are also pastes that contain both components. In all of these cases the adhesive is ready for use.

Nothing further need be added and no mixing is required. The curing agent used is latent and only begins to react with the epoxide resin at elevated temperatures. Hence the one-component types all require heat for curing.

There are of course, some limitations to the use of epoxide resin adhesives. Some materials such as certain crystalline or less polar plastics (e.g., polythene, silicones, and fluorocarbons) are not bonded effectively with epoxides. Processing epoxide systems can sometimes be difficult and careful surface preparation of the substrate is essential to realise the full potential strength of the adhesives. A rather rigid glue line is formed by the unflexibilised systems, which give high tensile and shear strengths but only low peel strengths. Careful joint design is therefore required to minimise 'peeling' forces.

The advantages of epoxide resin adhesives may be summarised thus:

(i) Their versatility in bonding a wide range of materials with flexible or rigid bonds, which retain their strengths from $-50\,°C$ to $+300\,°C$.

(ii) The extremely high bond strengths that can be achieved.

(iii) Low temperature curing, often at room temperature, accompanied by low pressure, often only contact pressure.

(iv) No volatile solvent is present and no volatiles are evolved during cure, which allows impervious surfaces to be bonded.

(v) Very small shrinkage occurs during cure, and hence less strain is built into the glue line.

(vi) The bonds formed show only low creep under prolonged stress and also have good heat resistance, electrical insulation, and resistance to corrosion, moisture, and many chemicals.

11.2 THEORY OF ADHESION

Several workers have demonstrated that the strength of an adhesive bond never approaches the theoretically possible values calculated from consideration of the molecular forces involved. De Bruyne,[4] for example, has shown that in the absence of an adhesive, the bonding forces alone between two metal discs 4 Å apart should be 68 000 lb/in². Taylor and Rutzler,[5] in studying the spatial fit between different polymers on various substrates calculated the force of adhesion based on the maximum and minimum number of dipole-ion interactions between adhesive and adherend.

In both cases the minimum calculated force of adhesion was considerably greater than the measured experimental values.

As Wake[6] states, 'if the cohesion of matter is adequately explained by interatomic and intermolecular forces why should not different sorts of matter adhere to each other with the same strength as the two sorts separately cohere?' It can therefore be concluded that only a small proportion of the maximum theoretical number of interactions between adhesive and adherent is occurring in practice. This failure to obtain a high level of interaction probably leads to flaws and cracks at the interface which serve as initiatory points for subsequent fracture propagation and ultimately bond failure. This mechanism of adhesion failure parallels that for the cohesive failure of materials, also thought to be due to structure imperfections and the propagation of fracture from flaws and cracks.

The foregoing considerations have assumed that the adsorption theory of adhesion is the correct one. This postulates that the adhesive is adsorbed either via formation of covalent bonds at active sites on the adsorbent surface or via van der Waals forces reinforced by dipole-dipole interaction.

An alternative theory developed chiefly in the Soviet Union by Deryagin and his co-workers, and described by Voyutskii,[7] proposes that hydrogen bond formation between adherend and adhesive leads to the formation of an electrical double layer which accounts for the bond strength. For a fuller consideration of the various theories of adhesion the reader is referred to the review by Wake.[6]

One factor which can cause failure is inadequate wetting of adherend by adhesive. This leads to incomplete contact, or, as shown by De Bruyne,[8] to development of stress concentrations when the adhesive exhibits a high contact angle on the adherend. Differences in thermal expansion coefficients between adhesive and adherend can also lead to internal stresses. Thus, May and Nixon[9] suggest that room temperature bond strength is dependent upon glass transition temperature (T_g). Above T_g, the elastomeric state of the polymeric adhesive would accommodate stresses developed in the system. Below T_g the rigid adhesive could not accommodate the stresses so easily and the higher stress levels therefore present lead to lower bond strengths.

The adhesive properties of epoxide resins are probably due to the pendant secondary hydroxyl groups which occur along the molecular chain and which are strongly adsorbed on to oxide and hydroxyl surfaces. Glazer[10] studied the collapse pressure of a series of epoxides deposited as monolayers on water and found that the ether oxygens and hydroxyl groups were orientated towards the water layer and that the epoxide groups were not adding to the adhesion to the water but rather assisted the lateral cohesion of the film. De Bruyne[11] followed this work by measuring the bond strengths of a series of different molecular weight resins based on DPP and resorcinol, and cured with phthalic anhydride. He found that the lap shear strength of the bonds was a function of the concentration of hydroxyl groups in the resin. An even better correlation was found when the number of hydroxyl groups per unit area of surface was considered, and the bond strength plotted as a function of (volume concentration of hydroxyl groups)$^{2/3}$. De Bruyne also recognised that factors other than hydroxyl content and molecular weight were varying in the resin systems. The stress set up by shrinkage on cure would also decrease with decreasing cross-link density, which is proportional to increasing hydroxyl content.

11.3 ADHESIVE FORMULATION

The low molecular weight diglycidyl ether liquid resin forms the basis for most formulations since it provides the best balance between ease of handling and properties of cured adhesive. Resins containing reactive diluents may also be used, their lower viscosity assisting in wetting of the substrate, allowing more filler to be used, and improving handling properties. These advantages are accompanied by a decrease in strength in the final adhesive. May and Nixon[12] report the reduction of hot strength to one-sixth of its room temperature value in formulations using AGE as a monoepoxide diluent. However, some diepoxide diluents examined imparted hot strengths that were greater than for the undiluted system.

The higher molecular weight resins are also used, e.g., the semi-solid resin WPE 225–280 and the solid resin (melting point 65–75°C) WPE 450–500. These grades improve the strength properties of the adhesives, which is probably due to the increased number of hydroxyl groups in the molecular chain. Improved high temperature performance and chemical resistance is attained by epoxidised novolak resins which because of their high viscosity are most commonly used in tape-supported adhesives. The cycloaliphatic resins do not appear at present to offer any improvement in adhesive performance over the conventional diglycidyl ether resins, at least at room temperature.

The curing of an epoxide resin for adhesive use is no different to that in any other application. Nevertheless, some curing agents are especially good for bonding specific materials, DICY being valuable for certain metals. The amounts of curing agent used can be a large proportion of the total mass of the reactants, so that the curing agent makes a significant contribution on its own behalf to the physical properties of the final cross-linked system. In addition to the structure, reactivity, and functionality of both resin and curing agents, the temperature of curing also plays an important part in determining the strength of the adhesive bond.

For hardening at room temperature or slightly elevated temperatures, the alkylene polyamines, are often selected, especially DTA, TET, and DEAPA. The order in which they are written above is also their decreasing order of reactivity, and DEAPA really requires heat-curing to achieve acceptable results. Whilst all form adhesives giving good bond-strengths with a wide variety of materials after

Table 11.1 EFFECT OF TEMPERATURE ON BOND STRENGTH[*13]

Curing agent	Cure temperature (°C)	Cure time	Bond strength (lb/in^2)
Triethylene-tetramine	25	3 days	1 162
	25	15 days	1 690
	95	30 min	3 172
	145	30 min	3 426
N,N-diethyl-aminopropyl-amine	40	16 hr	702
	40+25	16 hr + 14 days	840
	95	5 hr	3 236
	145	30 min	4 056

*$\frac{1}{2}$ in overlaps 0·064 in 24 ST–3 aluminium alloy (MIL-A-005090E)

room-temperature cure, hot curing or post-curing at 80–100 °C increases bond strengths significantly. This is well illustrated by the data (Table 11.1) given by Houwink and Salomon[13] for curing one particular resin grade by TET and DEAPA.

Other important curing agents suitable for cures at room temperature or slightly elevated temperature are the polyamide resins. Since they have high molecular weights in proportion to their amine content they are used in large amounts for curing purposes, and the mixing proportion with resins is not very critical. They have the added advantage of being essentially non-hygroscopic and non-irritant. These factors make them ideal hardeners for two-component epoxide resin adhesives sold for domestic use. The bond strengths of polyamide-cured adhesives also improve with increasing cure temperature, in one case rising from about 1 750–2 250 lb/in^2 after room temperature cure to 5 700–6 000 lb/in^2 after curing at 145 °C for 30 min. Polyamides are also used to improve the flexibility of the cross-linked polymer, an effect which improves peel strengths. However, it is usually accompanied by a decrease in the hot strength of the bond.

The aromatic polyamines and the acid anhydride hardeners may also be used in hot-curing adhesive formulations, but there is a dearth of published data on the performance of these systems, Houwink and Salomon[13] quoting from unpublished data by Ciba Ltd.

Boron trifluoride-amine complexes and DICY can both be regarded as latent catalysts, being non-reactive with the resins at room temperature, though effecting cure when the temperature is raised to 95 °C for the former curing agent and 180 °C for the latter. The use of DICY in adhesive formulation is well established, especially for metal bonding, and forms the basis for one-component adhesives in the form of rods or powders. To prepare these systems the DICY is dispersed in a liquid resin by grinding in a ball mill, or milled into a solid resin. The resulting mixture, which can be melted and cast into rods or left as a powder, has a shelf-life of about six months and cures rapidly at temperatures of 180 °C.

Unmodified epoxide adhesives are rather rigid when cured and do not show very good peel- and bond-strengths. Modification of the formulation can lead to a marked improvement in flexibility and toughness. One method is the incorporation of up to 10% by weight of a high molecular weight epoxide resin. This leads to improved flexibility and heat resistance, but a reactive diluent frequently needs to be used to counteract the increasing viscosity caused by this addition.

The addition of polysulphide rubbers also greatly improves peel and tensile shear strengths at room temperature, but the high temperature resistance of these systems is poor (Table 11.2).

Impact and peel strengths can also be improved, but hot strengths decreased, by the incorporation of low-M grades of poly(vinyl formal), poly(vinyl acetate), or poly(vinyl butyral). These are highly viscous materials, not easily mixed with the resin, and are usually used at concentrations of 10–20% maximum.

Fillers are often used in epoxide adhesive formulations to reduce

Table 11.2 EFFECT OF TEMPERATURE ON FLEXIBILISED ADHESIVES[14]

Curing agent (phr)	Flexibiliser (phr)	Cure schedule (min/degC)	Tensile shear strength (lb/in²)		
			−57°C	25°C	80°C
DMP–30 (10)	Polysulphide (50)	90/90	2 740	4 350	820
DTA (11)	Polysulphide (50)	90/90	3 050	3 820	630
Polyamide (120)	—	30/150	3 180	3 970	750

The same resin grade (low mol. wt. liquid diglycidyl ether) together with 30 phr.
Asbestos filler used throughout

shrinkage, thermal expansion, and cost, and in certain cases to increase bond strengths by improving the stress distribution in the glue line. They also change the physical state of the adhesive, making it thicker and sometimes imparting a degree of thixotropy, which is frequently desirable. Young[15] reports significant increases in strength on the incorporation of small amounts of finely divided silica. Asbestos fibres at 20% concentration are reported by Elam[16] to increase the Izod impact strength of an epoxy adhesive from 5·2 to 18 ft–lb/in².

11.4 JOINT DESIGN

Proper joint design is essential if the high strength of epoxide resin adhesives is to be fully utilised. Four basic modes of stressing in joints can be defined (Fig. 11.1).

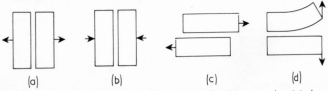

(a)   (b)   (c)   (d)

Fig. 11.1 Four basic modes of stressing in joints: (a) tensile, (b) compressive, (c) shear, and (d) peel

Of these, peel stress is the most critical in joint design. In the action of peeling, the adhesive bond is subjected simultaneously to two separate types of stressing force, tension and shear, and the stresses are concentrated along the zone of contact of substrate to adhesive.

If the substrate is thin and flexible the contact area becomes very small and the local tensile stress extremely high. Epoxide resin adhesives and rigid adhesives generally are poor in peel strength, and resin formulation plus joint design have been developed to minimise peeling forces.

When used in a lap joint under tensile shear, a flexible adhesive is more able than a rigid adhesive to distribute the peeling stresses over the bonding area. The strength of an adhesive joint is therefore greatly dependent upon the flexibility of the adhesive, and of the surfaces to be bonded, in addition to the magnitude of the peeling stress applied to the joint as part of the overall stressing.

Scarf or reinforced butt joints (Fig. 11.2) help to minimise peel

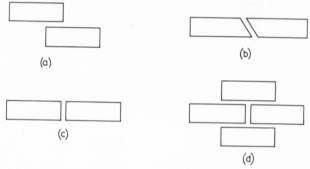

Fig. 11.2 *Scarf or reinforced butt joints: (a) simple lap joint, (b) scarf joint, (c) butt joint, and (d) double-strapped butt joint*

stresses and are simple and effective ways of improving the joint strengths. A review of joints and the stresses therein is given by Wake.[6]

11.5 SURFACE PREPARATION

All adhesive processes start with pretreatment of the surfaces to be bonded. This is not merely a cleaning or roughening, to remove gross surface contamination; it is often to change the chemical nature of the surface so as to render it more chemically reactive towards the adhesive molecules, to increase the 'wettability' of the surface, or to replace a weak layer of surface molecules by a more intact and coherent layer. Thelen[17] lists eleven reasons for preparing the substrate surfaces before the adhesive is applied. Each adhesive/

adherend system has its own specific surface treatment, which is found empirically to yield the best results.

In general, wood and other porous materials should be sanded and free from dust. Non-porous non-metallics should be degreased with detergent or solvents, rinsed with water, and dried. Metals are usually vapour-degreased with trichloroethylene, followed by acid etching or sand blasting; the last-named must be followed by further degreasing.

Plastics are degreased, roughened, and washed, followed (in the case of polyethylene and polypropylene) by an oxidative treatment using chromic acid or an oxygen-rich flame. Glass, decorative laminates, perspex, etc., are best degreased and abraded.

An exhaustive account of the recommended surface treatments for epoxide resins bonding is given by CIBA (ARL) Ltd.[18]

11.6 SOME BASIC ADHESIVE FORMULATIONS

Two suitable light duty general purpose adhesives, curing at room temperature, are given in Table 11.3.

Table 11.3 TYPICAL FORMULATIONS FOR A GENERAL PURPOSE EPOXIDE ADHESIVE

	I	*II*
	Parts by wt.	*Parts by wt.*
Low-*M* liquid diglycidyl ether resin	100	100
Talc filler	80	—
Tabular aluminium	—	50
DTA	11	—
Polyamide	—	80–100

These systems do not exhibit high strengths, but are easily handled and are adequate for many industrial applications even when cured at room temperature. The polyamide system is particularly suitable for domestic use since the mixing ratio of polyamide to resin is not critical; the polyamide is not unpleasant to handle, and is non-irritant.

Adhesive systems that are tough and flexible are given in Table 11.2; these can be cured at room temperature, but here maximum strength is not developed. Adhesive formulations of high peel strength are given in Table 11.4, together with some strength data.

Table 11.4 FORMULATIONS FOR HIGH PEEL STRENGTH EPOXIDE ADHESIVES

	III	IV
Low molecular weight liquid diglycidyl ether resin	100 parts by wt.	100 parts by wt.
Asbestos filler	30 phr	30 phr
Poly(vinyl acetate)	10 phr	10 phr
DEAPA	8 phr	—
DICY	—	6 phr
Cure schedule	90 min at 90 °C	90 min at 175 °C
Tensile shear strength		
(lb/in²) at 57 °C	2 650	3 200
(lb/in²) at 25 °C	3 740	3 320
(lb/in²) at 80 °C	3 635	4 170
(lb/in²) at 120 °C	1 180	2 660
(lb/in²) at 150 °C	530	1 325

When adhesives are needed for extreme temperature applications, epoxide resins may be used in conjunction with phenolic resins. Skeist[19] describes the Shell Adhesive 422 system which is a glass tape impregnated with an adhesive composed of 33 parts Epon 1 001, 67 parts Plyothen 5 023, 6 parts DICY, and 100 parts aluminium dust. Other epoxide–phenolic adhesives are also described which give good ageing performance at 288 °C. Other approaches to the high-temperature adhesive problem are (*a*) the use of a high temperature resistant curing agent such as PMDA, or (*b*) the use of polyfunctional epoxides such as epoxidised novolak resins. Both systems are discussed by Brennan, Lum, and Riley.[20]

Kausen[21] has reviewed the use of epoxide adhesives both at high and cryogenic temperatures, and combinations of epoxides with low molecular weight nylons show consistently high strengths at low temperatures.

REFERENCES

1. DAVIS, W. L. and PINKSTAFF, E., *Proc. Am. Soc. Civ. Engrs.*, **86**, CO1, 55 (1960)
2. FARR, F., *Brick, Clay Record*, **131**, No. 1, 48 (1957)
3. GAUL, R. W. and APTON, A. J., *Civil Engng.*, **29**, No. 11, 50 (1959)
4. DE BRUYNE, N. A., *Trans. Plast. Inst., Lond.*, **27**, 140 (1959)
5. TAYLOR, D. and RUTZLER, J. E., *Ind. Engng. Chem.*, **50**, 928 (1958)
6. WAKE, W. C., Adhesives, Lecture Series 1966, No. 4, The Royal Institute of Chemistry, London
7. VOYUTSKII, S. S., *Adhes. Age*, **5**, 4, 30 (1962)
8. DE BRUYNE, N. A., *Research, Lond.*, **6**, 362 (1953)
9. MAY, C. A. and NIXON, A. C., *J. Chem. Engng.*, **6**, No. 2, 290 (1961)
10. GLAZER, J., *J. Polym. Sci.*, **13**, 355 (1954)

11. DE BRUYNE, N. A., *J. appl. Chem., Lond.*, **6,** 303 (1956)
12. MAY, C. A. and NIXON, A. C., *Ind. Engng. Chem.*, **53,** 303 (1961)
13. HOUWINK, R. and SALOMON, G., *Adhesion and Adhesives*, Elsevier, Amsterdam, 2nd Ed. (1965)
14. Shell Chemical Co., Technical Literature
15. YOUNG, L. O., *Adhes. Age*, **2,** 11, 26, (1959)
16. ELAM, D. W., *Product Engng.*, **25,** 7, 166 (1954)
17. THELEN, E., *Handbook of Adhesives*, Ed. I. Skeist, Reinhold, New York, Chap. 3, 43 (1962)
18. Ciba (ARL) Ltd., Technical Literature
19. SKEIST, I., *Epoxy Resins*, Reinhold, New York (1958)
20. BRENNAN, W., LUM, D. and RILEY, M. W., *High Temperature Plastics*, Reinhold, New York (1962)
21. KAUSEN, R. C., *Mater. Des. Engng.*, **60,** 2, 94 (1964) and **60,** 3, 108 (1964)

BIBLIOGRAPHY

Adhesion, Ed. ELEY, D. D., Oxford (1961)
The Science of Adhesive Joints, BIKERMAN, J. J., Academic Press, New York (1968)
Adhesion and Adhesives, HOUWINK, R. and SALOMON, G., Elsevier, Amsterdam, 2nd Ed. (1965)

CHAPTER 12

Epoxide Resins in Building and Civil Engineering

The extreme versatility of epoxide resin formulations has led to their widespread use in the constructional industries of many countries. The simple mixing and polymerisation methods involved enable them to be used easily 'on site', and their extremely high bond strengths to many different substrates, coupled with their outstanding strength and chemical resistance, have suggested uses in a variety of building operations.

12.1 FLOORING

One of the chief uses is for screeded or self-levelling industrial flooring. In certain situations, floors based on conventional materials such as cement concrete break down very rapidly, so that floor maintenance is a severe problem. Chemical spillage, heavy wear by pedestrians and vehicles, and vigorous cleaning are all factors that can lead to the disintegration of a floor; and these conditions are widely found in chemical plants, refineries, food factories, plating and pickling shops, canneries, breweries, and warehouses.

Epoxide resin floors consist essentially of the binder resin, curing agent, any other ancillary chemicals in the formulation, and the aggregate (i.e., filler), plus pigments if required. They are usually applied to concrete, metal, or wood substrates, and a range of colours, surface finishes, and thicknesses is possible. Important characteristics of these floors are:

(a) excellent adhesion to a variety of surfaces,
(b) outstanding chemical resistance,

222

(c) high tensile, compressive, impact, and flexural strengths,

(d) easy application and rapid cure,

(e) lightweight compared with concrete, and

(f) jointless, dustfree, skid-resistant, and readily cleaned.

For all epoxide flooring applications the proper preparation of the substrate is of crucial importance to the success of the floor and the attainment of maximum performance. For concrete a clean, sound, powder-free surface is required, and this can generally be obtained by wire-brushing, sand-blasting, or acid-etching. Steel must be degreased, and any rust, scale, or heavy soiling removed by wire-brushing or other mechanical means. After cleaning, a tack or priming coat of unfilled binder system is brushed or sprayed on to the surface to ensure good adhesion of the filled composition to the sub-floor, and to seal small cracks and wet the substrate, so preventing air bubble formation, particularly in the self-levelling system.

12.1.1 SCREEDED EPOXIDE FLOORING

In this system the resin binder is usually filled with graded silica sand or calcined bauxite at a loading of about 85% filler. The bauxite confers a wear resistance several times greater than that of Portland cement screeds. Typical formulations for trowelled applications are:[1]

	Parts by wt.	*Parts by wt.*
Binder		
Liquid diglycidyl ether resin	100	100
Pine oil (flexibiliser)	27	20
Phenol (cure accelerator)	5	5
DTA	13	17
Coal tar (extender)	—	110
Aggregate		
Silica sand or calcined bauxite at an aggregate-to-binder ratio of 7:1		

The aggregate must be dry. It must also have the correct particle size distribution, which sometimes requires the blending of two or more aggregates. A maximum void content of 25% is needed, in contrast to commercially available sands which usually have void contents between 35 and 40%.

Resin, curing agent, and pigment paste are thoroughly mixed, preferably mechanically, and the aggregate added. Mixing is continued for 3–5 min and the mixture dumped on to the floor and spread out evenly. Spreading out increases the surface-to-mass ratio, and hence decreases exothermic heat build-up and lengthens working life of the mixture.

The roughly screeded mix, usually $\frac{1}{8}$–$\frac{1}{4}$ in thick, is compacted or tamped to densify the floor and finally smoothed to a finish by trowel. The floor will be sufficiently hard to withstand normal usage after 24 hr and will attain full chemical resistance and strength after 7 days. This speed of hardening enables savings to be made in avoiding excessive downtime for floor repairs or maintenance.

Typical mechanical properties of trowelled epoxide resin floors compared with concrete are:[2]

Property	Epoxide system	Concrete
Ultimate tensile strength, lb/in²	1–2 000	200–500
Ultimate compressive strength, lb/in²	8–12 000	3–6 000
Impact resistance, ft–lb	10–20	11
Ultimate flexural strength, lb/in²	1–4 000	1 000
Specific gravity	1·9–2·1	—

Attractive epoxide resin terrazzo floors are also possible in both tile and jointless forms,[3] the resin replacing the conventional cement binder. After hardening, the trowelled floor containing marble chips can be ground to a decorative finish in the normal way. Epoxide terrazzo has the advantage of not requiring a thick under-bed, and a significant weight saving is hence achieved. It can also be ready for use much quicker than the cement terrazzo floor.

Large areas of trowelled epoxide floors have been laid on sites such as warehouses, factories, laboratories, power stations, and hospitals, the terrazzo type being found in banks, student halls of residence, and office foyers.

12.1.2 SELF-LEVELLING EPOXIDE FLOORING

These formulations have a low viscosity and filler content, both ensuring a free-flowing mixture. They are poured on to concrete or other rigid substrate and spread out with a plastic comb, squeegee, or broom to a thickness of about $\frac{1}{16}$ in. The substrate must therefore be even enough to accept this thickness.

A typical formulation would be:

	Parts by wt.
Liquid diglycidyl ether resin	300
Furfuryl alcohol	18
Triphenyl phosphite	24
Laromin C 252	13
Laromin C 260	15
Furfuryl alcohol	10

This binder is used with a chemically inert filler such as quartz powder or calcined bauxite at a filler-to-binder ratio of 7:4 or 9:4 respectively. The particle size distribution of the filler is important and is usually in the range 140–250 mesh ASTM, plus some fines smaller than 200 mesh. Particle shape and density is also important.

The mixed binder and filler combination is spread uniformly over the surface of the primer coat. If non-skid properties are needed, carborundum granules can be lightly broadcast over the surface of the wet floor, and these sink into the floor until they can hardly be felt. After the floor has hardened (usually 24 hr at 20 °C) it can be polished to remove dirt, producing an attractive glossy non-skid surface. Alternatively, a matt finish can be produced.

The self-levelling system has wear characteristics at least as good as the highest quality PVC or linoleum and has the added advantage of withstanding indentation by such things as stiletto heels. It is not suitable for areas where heavy abrasive wear is encountered, but is widely used for continuous floors in hospitals, schools, kitchens, light industrial factories, and numerous other situations.

12.2 ROAD AND BRIDGE COATINGS

The application of epoxide resin systems to roads and bridges first occurred in the U.S.A. They are used as a protective membrane to prevent spalling of concrete road and bridge surfaces, and to overcome the ingress of water via cracks in the concrete of bridge decking which subsequently corrodes the steel structure underneath. In addition the epoxide surfacing can be given non-skid properties by using an appropriate filler.[4, 5] These thin surfacings also give protection to the concrete from de-icing salts, fuels, and lubricants.

The formulations frequently contain liquid epoxide resins cured with DTA or other amine, together with a high proportion of coal tar as an extender. This is applied in a surface dressing technique by a mobile spraying system or by hand, followed by broadcasting the aggregate on to the wet surface, providing a topping of about $\frac{1}{16}$ in thick. Wittenwyler[6] has given a useful review of the situation in the U.S.A. and there are now many articles demonstrating the use of this technique.[7]

In the United Kingdom an assessment of epoxide resin systems for road and bridge surfacings has been made by James[8] who classified the possible applications as:

(a) thin coatings for concrete roads that are scaling,
(b) thin coatings for asphalt or concrete surfaces which have become slippery but are still sound,
(c) surfacings for sites where fuel spillage is a problem,
(d) thin lightweight surfacings for bridge decks,
(e) road marking materials,
(f) coloured surfacings, and
(g) additions to conventional bituminous surfacings to improve the performance.

12.3 CONCRETE BONDING AND REPAIR

Using their extremely good adhesive and mechanical strength properties, filled epoxide resin compositions are ideal for the repair of cracks in concrete, wood, brick, metal, and many other materials.[9] An important feature is their ability to form strong bonds between wet alkaline materials such as concrete.

For grouting or crack repair the filled adhesive is poured, brushed, pumped under pressure, or sometimes trowelled into the crack when the gap is sufficiently wide. The strength and slight flexibility of the epoxide binder are sufficient to resist load and temperature stresses and prevent progressive concrete failure that would otherwise occur at cracks. Concrete dams, bridges, piers, roads, and paths have all been easily and effectively repaired by this technique. This application has already been briefly mentioned in Chapter 11, together with many other important uses for epoxide adhesives in the constructional industry.

12.4 SOIL CONSOLIDATION

An ingenious use for epoxide resins is in oil wells drilled into geological formations of loose sand. The sand can enter the well and cause blockages, hence reducing the output. This difficulty has been successfully overcome by using a solvent-containing amine-cured system to consolidate the sand around the well bore. The consolidated sand then becomes a filter, which prevents further movement of sand into the well itself. The process of consolidation first requires the water in a small area of the sand round the bottom of the well to be removed by an alcoholic solvent. The liquid resin system is then pumped into the pore space of the loose sand. As cure proceeds, a liquid polymer first separates out and spreads over the sand grain surfaces, concentrating at grain-to-grain contact points. This liquid phase on undergoing further cure forms the usual tough cross-linked polymer which cements the grains firmly together, giving a consolidated but permeable structure of considerably increased strength. A full description of the process is given by Havenaar and Meijs.[10]

A similar application to sand consolidation is the use of an epoxide resin formulation to seal off part of a Ruhr coal mine shaft to prevent the ingress of water containing a high concentration of mineral salts.[11]

12.5 MISCELLANEOUS USES

Lightweight exterior cladding panels for buildings have been coated with an epoxide system, followed by an aggregate to protect and enhance the appearance of the panelling.[12] Epoxide resin mortars have been used as tile grouts which have outstanding chemical resistance. Kitchen sinks and shower bases have been fabricated from epoxide moulding powders by a compression moulding technique. Reflecting studs for road use have been made from reflecting particles (small glass spheres) dispersed in the epoxide resin system, the stud being bonded to the road by an epoxide adhesive. Epoxide resin laminates have been used as shuttering for structural concrete and for the preparation of decorative concrete facing panels. There are numerous other examples of epoxide resin compositions being used in the constructional industry and a RILEM Symposium[13] on the use of synthetic resins in the industry gives details of new applications that are being considered. It is clear that the present usage represents only the beginning of a much greater exploitation of these highly versatile resins in that industry.

REFERENCES

1. Shell Chemical Co., Technical Literature
2. Ciba (ARL) Ltd., Technical Literature
3. ANON., *Bldg. Constr.*, **33,** No. 3, 58 (1963)
4. NAGIN, H. S., NOCK, T. G. and WITTENWYLER, C. V., *Bull. Highw. Res. Bd.,* Washington D.C., No. 184, paper 1 (1957)
5. CREAMER, W. M. and BROWN, R. E., *Bull. Highw. Res. Bd.,* Washington D.C., No. 184, paper 2 (1957)
6. WITTENWYLER, C. V., *Highw. Res. Abstr.* (1962)
7. MACKENZIE, T. T., *Civil Engng.*, **32,** No. 3, 42 (1962)
   MINARIK, W. L., *Pub. Wks.*, **94,** No. 7, 97 (1963)
   ANON., *Contractors and Engineers,* **60,** No. 9, 22 (1963)
   ANON., *The American City,* **79,** No. 1 (1964)
   ANON., *Contractors and Engineers,* **61,** No. 9, 58 (1964)
8. JAMES, J. G., *Rds. Rd. Constr.*, **41,** 236, 488 (1963)
9. GAUL, R. W. and APTON, A. J., *Civil Engng.*, **29,** No. 11, 50 (1959)
10. HAVENAAR, I. and MEIJS, F. H., *J. Instn. Petrol.*, **49,** 480, 382 (1963)
11. AU, E., *Min. Mag. Lond.*, **171,** No. 3 (1964)
12. ANON., *Mod. Plast.*, **42,** No. 7, 116 (1965)
13. International Union of Testing and Research Laboratories for Materials and Structures, *International Symposium, Paris* (1967)

CHAPTER 13

Miscellaneous Applications

13.1 TOOLING

One of the early practical applications of epoxide resins was their use in the aircraft and motorcar industries for tooling. Subsequently many other industries realised the advantages to be gained by using the resins in this way, and diverse industries such as ship and boat building, foundry, domestic equipment, and pottery all now employ epoxide resins in their production lines.

The term 'tooling' as used here embraces the following items:

(a) prototype and master models for product design,

(b) drilling and welding jigs, checking fixtures,

(c) moulds used as masters for die-sinking and Keller models; moulds for the vacuum forming and injection moulding of thermoplastics; moulds for the explosive forming of metals,

(d) foundry patterns, and

(e) drop-hammer and press-tools; stretch blocks.

In all of these examples, the use of the resins has not introduced any new basic approach to engineering practice, but rather has brought about large savings in time and money, and provided a tool which often has a performance superior to that of one made in conventional materials.

The manufacture of resin tools uses simple casting and hand lay-up laminating techniques. The characteristics which have led to their widespread adoption are chiefly:

(a) the ease with which complicated shapes and complex con-toured surfaces can be formed by simple casting techniques, without the need for subsequent machining, (*b*) the lighter weight of a resin as compared with a metal tool, and

(*c*) resistance to corrosion and inertness towards changes in atmospheric conditions which would alter the size and accuracy of a wooden tool or pattern.

Valuable detailed treatments of this subject are available from company literature[1] and also the American Society of Tool and Manufacturing Engineers monograph.[2]

13.1.1 MODEL MAKING

In the development of new models of motor cars, a clay model is usually made from which a plaster female mould is cast. This negative mould is then used to construct an epoxide resin master mould by lining the plaster mould with a thin layer of micro-balloon-filled epoxide resins, and backing this shell with a glass/epoxide laminate. The whole structure is braced with laminated epoxide resin tubing. This resin master, whose surface can be cut and shaped like wood, can be used for making other masters by the same process which will be used for subsequent design of tools etc.

In a similar way, large die-sinking epoxide Keller models can be manufactured. These models are used to guide a Keller cutting machine, which is similar to a pantograph. The profile of the epoxide model is traced by the machine and the cutter of the machine simultaneously produces the same shape in tool steel.

13.1.2 JIGS AND FIXTURES

These tools are used as aids to production, assembly, and inspection. Depending on their shape, size, and end-use they can be made as solid castings or as braced laminates. Both types are strong, light in weight, and able to withstand rough handling, and to retain their accuracy even after a heavy blow.

Jigs are used to ensure rapid production by accurately positioning components with respect to other parts and hence assisting drilling, cutting, welding, etc., to be carried out in a closely controlled and rapid way. Fixtures are a means of quickly checking the shape, position, and dimensions of a component.

13.1.3 MOULDS

Epoxide resin moulds, prepared from a clay, wood, or plaster master by the laminating technique, are widely used for polyester laminate production by the contact pressure process. The master is polished smooth, and treated with a release agent and then an epoxide resin gel coat. This is followed by alternate layers of glass cloth and resin in the usual hand lay-up method (Chapter 10). The laminate is then braced with a framework of epoxide glass tubing of about 1–2 in diameter.

Matched moulds for dough moulding and the production of epoxide or polyester laminated articles by the pressure moulding process, can also be cast from epoxide resins. These moulds are cheaper and much easier to construct than metal ones, especially when the mould surface has a complex shape.

Vacuum-forming of thermoplastics such as polythene, polypropylene, PVC, polystyrene, and cellulose acetate can be carried out by using epoxide resin moulds. The preheated plastic sheet is drawn on to the epoxide mould by a vacuum applied through perforations in the latter, the sheet assuming the exact shape of the mould profile.

The chief advantage in using resin moulds in this application is the ease with which they are cast to true size, thus eliminating expensive machining and finishing. Modifications to the tool are also readily carried out, which is not always possible with an aluminium alloy mould.

The construction in epoxide resins of injection moulds for thermoplastics is of advantage where a short production run would be exceedingly expensive if metal moulds were to be made, or where prototype components are being made for market evaluation or the possibility exists that the design may be changed to meet customer demand.

13.1.4 FOUNDRY PATTERNS

Epoxide resins are widely used in foundries for the construction of master moulds from the original pattern, for duplicate patterns made from the master mould, and for core boxes. The reasons for this widespread use of resins lies in their toughness, dimensional stability, and ability to withstand rough handling when in cast or laminated form, plus their fidelity in reproducing the shape of a model.

Small patterns and core boxes are usually solid castings. For large castings, considerable savings in cost and reduction in exothermic heat evolution can be achieved by casting a facing resin system around a core of highly filled epoxide resin. The core is first prepared by lining the mould surface with a sandwich of polythene sheets containing a layer of modelling clay to occupy the thickness of the facing resin. The core mix, which is highly filled with such materials as sand, vermiculite, marble flour, or broken pieces of cured resin, is tamped into position and allowed to cure. It is then removed, its surface abraded, the mould surface treated with release agent, and the core repositioned in the mould. The facing layer of resin is then poured into the gap between core and the mould and allowed to cure.

13.1.5 TOOLS FOR METAL FORMING

Epoxide resin tools have been used for explosive forming, stretch blocks, rubber bed press tools, drop-hammer punches and dies, and dies for press forming. They are of particular value in prototype and short-to-medium production runs where it would be uneconomic to produce a high quality steel tool. Runs up to 10 000 units have been shown to be economical when using resin tools.

The construction of resin metal-forming tools such as punch and die sets usually employs a low-cost rigid core faced with a high quality resin system, as described in the previous section. Alternatively, a resin face can be cast on to an existing metal punch. The compressive strength, impact strength, and surface hardness of the resin system are clearly of great importance for these applications. However, in certain drop-hammer work it is necessary to provide a resilient facing resin on one of the tools of the set.

13.1.6 FORMULATIONS

Most tooling formulations are based on the liquid glycidyl ether resins, cured with polyalkylene polyamines for room temperature curing or modified aromatic amines or mixtures of them for hot-cured systems. Diluents, plasticisers, and flexibilisers are also employed when appropriate, and fillers are widely used to reduce cost and improve handling and physical properties.

There are numerous accounts in the technical press of the use of epoxide resins for tooling.[3]

13.2 EPOXIDE RESIN FOAMS[4, 5]

Methods for producing rigid foams from epoxide resins have been known for many years, and in the period 1959–62 much research effort in the U.S.A. was devoted to developing viable commercial systems. In the event this effort was fruitless, and to date most rigid foam applications use polyurethanes which offer comparable or better properties at lower cost. Nevertheless, it is of interest to consider briefly the methods of foam production. They can be conveniently classified into two categories, blown foams and syntactic foams.

Blown foams are formed when a gas is released in the system during cure. The gas can either be formed by the curing reaction itself, or from a compound present in the formulation which either evolves a gas or is vaporised itself by the heat of the polymerisation. An example of the first type is the trimethoxyboroxine-amine-epoxide system, where the boroxine curing agent reacts with the amine to form boron-nitrogen bonds with the liberation of methanol which acts as the blowing agent.

Hydrazide derivatives can be incorporated in the mix and decompose on heating to form nitrogen, which acts as a blowing agent. Low-boiling point liquids such as methanol or fluorotrichloro-methane (b.p. 23·8°C), will vaporise on heating and produce a foam.

Syntactic foams are epoxide resin systems filled with very small microballoons made from organic or inorganic materials. Organic microspheres are usually made from PF, UF, or polystyrene resins, and are filled with nitrogen or, in the case of polystyrene, a fluoro-carbon or pentane. Inorganic spheres can be of glass or aluminium silicate. In all except the polystyrene system, the epoxide polymerises and the microspheres remain unchanged and distributed through-out the rigid matrix. The polystyrene spheres, filled with volatile material, expand on heating to many times their original volume, heat being provided either from an external source or from the heat of polymerisation of the resin.

13.3 PVC STABILISATION

On exposure to heat and light, PVC breaks down with the formation of hydrogen chloride, which itself accelerates further decomposition. Combinations of epoxides and salts of barium, cadmium, strontium, and zirconium can stabilise PVC against this progressive degradation by removing the hydrogen chloride formed. The probable mechanism of the stabilisation effect is interaction of epoxide with hydrogen chloride, forming the chlorohydrin, followed by dehydrohalogenation by the metal salt, with regeneration of the epoxide group. The mixture of metal salt and epoxide resin acts synergistically, the stabilising effect being greater than that of the individual constituents.[6,7]

Along with a conventional stabiliser, the liquid diglycidyl ethers of DPP are effective in PVC stabilisation, but do not additionally help to plasticise the polymer. Other epoxides such as epoxidised oils, esters, and fatty acids[8-10] and some cycloaliphatic[11] and olefinic epoxides, are both good stabilisers and plasticisers and are now used in preference to the glycidyl ether resins.

13.4 APPLICATIONS OF PHENOXY RESINS

The phenoxy resins, described briefly in Section 2.1.7, have been developed as thermoplastic materials for blow moulding, injection moulding, extrusion moulding, and for adhesives and coatings. Blow moulding and extrusion grades have ideal characteristics for bottle manufacture—high clarity, toughness, rigidity, effective oxygen and moisture barrier, easy processability, low odour, and absence of taint.

Some physical properties of a bottle resin are:[12]

Tensile strength at yield, lb/in^2	9 000
Ultimate elongation, %	90
Flexural modulus, lb/in^2	410 000
Izod impact strength, ft–lb/in notch	2·5
Heat deflection temperature (66 lb/in^2 ; °C)	91
Mould shrinkage, in/in	0·004
Moisture vapour transmission ($g/100 in^2/10^{-3}$ in/day)	3–5

Other properties of the resin include better resistance to creep than most other thermoplastics, and only small changes in mechanical properties over a wide temperature range.

Phenoxy resins with a slightly lower molecular weight are suitable for adhesives and anticorrosive coatings. Since the 'phenoxies' are strongly polar polyethers, they are readily soluble in ketones, esters, and mixtures of aromatic hydrocarbons, chlorinated hydrocarbons, or alcohols with ketones.

13.5 EPOXIDE RESIN MOULDING POWDERS

As with other thermosetting resins (e.g., PF, UF, and MF), epoxide resins can be formulated as moulding powders which can be used in compression or transfer moulding techniques. The powders cure rapidly under the influence of heat and pressure without the evolution of volatiles and with negligible after shrinkage. The mouldings produced in general have high mechanical strength and chemical resistance, good electrical properties and heat resistance, and low water-absorption.

Table 13.1 TYPICAL PROPERTIES OF EPOXIDE RESIN MOULDINGS[12]

Property	Glass-filled powder	Mineral-filled powder
Specific gravity	1·73–1·78	1·76–1·94
Tensile strength, lb/in^2	9–11 500	7–10 000
Modulus of elasticity (tensile; lb/in^2)	1·4–2·5 × 10^6	1·1–1·7 × 10^6
Cross-breaking strength, lb/in^2	17–20 000	15–16 500
Impact strength, ft/lbf	0·6–3·0	0·2–0·45
Crushing strength, lb/in^2	20–29 000	23–27 000
Heat deflection temperature, °C	104–>200	100–125
Water absorption* (% weight increase; discs 5 in dia. × $\frac{1}{8}$ in thick, 24 hr at 20 °C)	0·02–0·03	0·01–0·02
Flammability*	self-extinguishing	self-extinguishing
Electrical strength (90 °C; 1 min V/10^{-3}in)	280–300	275–300
Dielectric constant, 50 Hz	5·0–5·8	4·9–5·4
10^3 Hz	4·9–5·2	4·9–5·2
10^6 Hz	4·8–5·2	4·9–5·3
Power factor (tan δ), 50 Hz	0·013–0·033	0·010–0·017
10^3 Hz	0·012–0·033	0·010–0·017
10^6 Hz	0·018–0·028	0·014–0·028
Volume resistivity, log$_{10}$ohm cm	>15	>15
Surface resistivity (after 24 hr in water; log$_{10}$ohm)	12–14	14–15
Tracking resistance*	excellent	excellent

All tests except those marked * were carried out as described in BS 2782

The use of epoxide resin moulding powders for the encapsulation of electrical components and the production of other items such as bobbins, has already been described (Section 9.6). The powders can also be used to produce chemical valves and filter plates and many other structural parts including cases and housings for the electrical and chemical industries.

Both solid and liquid glycidyl ether resins are employed in the powder formulations, often blended with an epoxidised novolak to obtain fast gel times as well as good flow properties. Aromatic amines such as DDM, or anhydrides (often PA, THPA, and HPA), are the usual curing agents. The remainder of the formulation consists of pigments, fillers, and a release agent. The effect of particulate fillers on the properties of the moulding powder is similar to their effect in liquid casting systems. Fibrous fillers such as glass, asbestos, or synthetic fibres, in lengths up to one inch, are also used and improve impact strength and other mechanical properties. Mould release is assisted by the inclusion of zinc stearate in the powder.

The typical physical and electrical properties possessed by epoxide mouldings incorporating glass fibres or particulate fillers are shown in Table 13.1.

REFERENCES

1. Ciba (ARL) Ltd., Technical Literature
2. *Plastics Tooling and Manufacturing Handbook,* published for American Society of Tool and Manufacturing Engineers, Prentice-Hall, New Jersey (1965)
3. DELMONTE, J., *Mod. Plast.,* **36,** No. 12, 82 (1959)
    DELMONTE, J., *Mater. and Methods,* **40,** No. 2, 93 (1954)
    DELMONTE, J., *Tool Eng.,* **37,** No. 1, 84 (1956)
    KISH, S. P., *Tool Eng.,* **36,** No. 1, 85 (1956)
    SOKOL, B., *Am. Mach.,* **100,** No. 5, 124 (1956)
    SPARROW, L. R., *Ind. Engng. Chem.,* **49,** No. 7, 1111 (1957)
    TIERNEY, J. W., *Foundry,* **85,** No. 12, 86 (1957)
4. FERRIGNO, T. H., Rigid Plastics Foam, 2nd Edition, Reinhold, New York, Chap. 3 (1967)
5. SCHNITZER, H. S. and RICHTER, S., *Mod. Plast.,* **37,** 2, 99 (1959)
6. WINKLER, D. E., *Ind. Engng. Chem.,* **50,** 2, 863 (1958)
7. DODGSON, D. P., *J. Oil Colour Chem. Ass.,* **43,** 576 (1960)
8. RISER, G. R., HUNTER, J. J., ARD, J. S. and WITNAUER, L. P., *S.P.E. Jl.,* **19,** No. 8, 729 (1963)
9. BRICE, R. M. and BUDDE, W. J., *Ind. Engng. Chem.,* **50,** No. 2, 868 (1958)
10. HENSCH, E. J. and WILBUR, A. G., *Ind. Engng. Chem.,* **50,** No. 2, 871 (1958)
11. VAN CLEEVE, R. and MULLINS, D. H., *Ind. Engng. Chem.,* **50,** No. 2, 873 (1958)
12. *Modern Plastics Encyclopedia,* McGraw-Hill, New York (1967)

Appendix

NOTE ON HANDLING EPOXIDE RESIN FORMULATIONS

Epoxide resins and associated chemicals, such as curing agents and diluents, should be handled with care. They are reactive materials and may therefore cause damage to human tissues. However, over the years the chief health hazard shown to exist is skin sensitisation and dermatitis.

The unmodified liquid diglycidyl ethers of DPP are not particularly irritating to the skin, but can nevertheless cause sensitisation in susceptible individuals. Resins containing a reactive diluent such as a mono-epoxide compound are distinctly more irritating and more likely to cause sensitisation. The solid resins are unlikely, normally, to cause any serious dermatitic response.

The amines (especially if aliphatic) comprise a more important source of dermatitic hazard. They can cause dermatitis and even burns on direct contact with the skin. In addition, some individuals can develop a sensitisation-type dermatitis at a first exposure, and at subsequent exposures the attacks become progressively more severe.

Aromatic amines are felt to offer less of a hazard from skin contact than the aliphatic types. Anhydrides can also cause skin irritation and even burns and some individuals are sensitised by them. Polyamides are distinctly less hazardous to handle.

The newer epoxides, such as the cycloaliphatics, would be expected to behave in a similar way to the reactive diluents, but as yet there is insufficient experience of their use to allow broad generalisations to be made. The low molecular weight diepoxides are known to be far more toxic than the diglycidyl ethers of DPP on ingestion.

In addition, they present a danger from skin penetration, and inhalation of their vapours must be avoided.

Finally, hazard from the cured resins is virtually non-existent. Only in grinding and machining operations, where dust is produced and possibly inhaled, is there some risk.

It is clear from these remarks that care must be taken when handling epoxide resin formulations. However, with properly thought-out procedures, the risk to health can be minimised, and the resins used quite satisfactorily. The basic conditions to be observed are:

(a) Avoid skin contact.

(b) Provide good ventilation and protective clothes such as rubber gloves and aprons.

(c) Practice personal hygiene.

The technical literature provided by the resin manufacturers contains useful and comprehensive guidance on the possible hazards involved and the precautions that should be taken when handling epoxide resin systems.

General Bibliography

BRUINS, P. F., Ed., *Epoxy Resin Technology,* Interscience, New York (1968)
LEE, H. and NEVILLE, K., *Handbook of Epoxy Resins,* McGraw-Hill, New York (1967)
PAQUIN, A. M., *Epoxyverbindungen und Epoxuharze,* Springer Verlag, Berlin (1958)
SKEIST, I., *Epoxy Resins,* Reinhold, New York (1958)

Index

241